TRANSPORT OF ANIMALS INTENDED FOR BREEDING, PRODUCTION AND SLAUGHTER

CURRENT TOPICS IN VETERINARY MEDICINE AND ANIMAL SCIENCE

VOLUME 18

Series ISBN: 90-247-2429-5

TRANSPORT OF ANIMALS INTENDED FOR BREEDING, PRODUCTION AND SLAUGHTER

A Seminar in the CEC Programme of
Coordination of Research on Animal Welfare,
organised by R. Moss, and held in
Brussels, 7-8 July, 1981

Sponsored by the Commission of the
European Communities, Directorate-General
for Agriculture, Coordination of Agricultural Research

Edited by

R. Moss
Ministry of Agriculture, Fisheries and Food,
London, United Kingdom

1982
MARTINUS NIJHOFF PUBLISHERS
THE HAGUE / BOSTON / LONDON

for

THE COMMISSION OF THE EUROPEAN COMMUNITIES

Distributors

for the United States and Canada
Kluwer Boston, Inc.
190 Old Derby Street
Hingham, MA 02043
USA

for all other countries
Kluwer Academic Publishers Group
Distribution Center
P.O. Box 322
3300 AH Dordrecht
The Netherlands

Library of Congress Cataloging in Publication Data
Main entry under title:

Transport of animals intended for breeding,
    production, and slaughter.

    (Current topics in veterinary medicine and
animal science ; v. 18)
    1. Livestock--Transportation--Congresses.
I. Moss, R.   II. Commission of the European
Communities.   Directorate-General for Agriculture.
III. Series.
SF89.T7 1982   636.08'3   82-7870
ISBN-13: 978-94-009-7584-2   AACR2
ISBN-13:978-94-009-7584-2   e-ISBN-13:978-94-009-7582-8
DOI: 10.1007/978-94-009-7582-8

Publication arranged by
Commission of the European Communities,
Directorate-General Information Market and Innovation

EUR 7913 EN

© ECSC, EEC, EAEC, Brussels-Luxemburg, 1982
Softcover reprint of the hardcover 1st edition 1982

Manuscript Preparation by
Janssen Services, 33a High Street, Chislehurst, Kent BR7 5AE, UK

LEGAL NOTICE

Neither the Commission of the European Communities nor any person acting on behalf of the Commission is responsible for the use which might be made of the following information.

CONTENTS

VI

PREFACE

The transport of farm livestock was the subject of the seminar held
from 7 - 8 July 1981 at the Commission of the European Communities (CEC),
Directorate General for Agriculture, Brussels as part of the work of the
Division Coordinating Agricultural Research. The aims of the seminar were
to examine the knowledge available on how the physiology and behaviour of
animals may change during transport; to consider the significance of these
changes in relation to welfare and economics and to assess those actions
which as experimental projects or observational studies might be proposed
to fill the most important gaps in our knowledge of the welfare of farm
animals during transport.

A number of conclusions can be drawn from the proceedings:

1. Much knowledge is available from both scientific observations and
    practical experience which could be used to improve the transport of
    livestock. Methods of loading, the construction of vehicles, ships,
    crates and aircraft could benefit from the application of existing
    knowledge. It is less clear whether it is best to concentrate on
    disseminating existing knowledge by education and advice or to
    contemplate more regulations.

2. Losses by down-grading at slaughter can largely be attributed to the
    ways in which animals are transported and handled.

3. There would appear to be considerable scope for further direct
    observation of livestock during transport, particularly since evid-
    ence is lacking on how animals prefer to align themselves or whether
    they prefer to stand or lie during a journey by road. There has been
    more observation during transport by air and sea.

4. A few areas are described where specific basic research is suggested
    to elucidate some key problems.

The proceedings of this seminar will help the CEC in relation to the
Council of the European Community Directive 81/389/EEC, which implements
Directive 77/489/EEC on the protection of animals during international
transport, although they do not complete the task of examining whether there
are good physiological, ethological and economic reasons for limiting the
final journey to the abattoir of animals intended for immediate slaughter
and, if so, on recommending a maximum duration for such journeys.

X

The CEC thanks the participants for their contributions and in also acknowledging the great help received in the organisation of the meeting hopes that the conclusions have stimulated some research efforts on this important subject in the Member States of the European Communities.

SESSION I

INTRODUCTION

Chairman: W. Sybesma

COUNCIL OF EUROPE CONVENTION ON TRANSPORT OF LIVESTOCK, EC COUNCIL OF MINISTERS DIRECTIVE ON PROTECTION OF ANIMALS DURING INTERNATIONAL TRANSPORT

H.C. Adler

Royal Veterinary and Agricultural University
Copenhagen, Denmark.

(1)   COUNCIL OF EUROPE CONVENTION ON TRANSPORT OF LIVESTOCK

The European Convention for the Protection of Animals during International Transport is dated 13th December 1968.   It was enforced on 20th February 1971.

It comprises domestic and non-domestic mammals, birds and cold-blooded animals in any movement crossing a frontier except frontier traffic.   The aim of the Convention is to establish common minimum provisions for the protection of animals ensuring their welfare.

For the purpose of this seminar the relevant animals are two groups of animal species:

1)   Domestic solipeds and domestic animals of the bovine, ovine, caprine and porcine species;

2)   domestic birds and rabbits.

Chapter I contains obligations of contracting parties and some definitions.   Chapter II in the Convention deals with 1), Chapter III with 2).

The ten member countries of the European Communities (EC) are all contracting parties of the convention and 8 non-member countries are contracting parties (Table 1).

TABLE 1

CONTRACTING PARTIES

EC countries	Non EC countries
Belgium	Austria
Denmark	Cyprus
France	Finland
Germany	Norway
Greece	Spain
Ireland	Sweden
Italy	Switzerland
Luxembourg	Turkey
Netherlands	
United Kingdom	

On 10th May 1979 the Additional Protocol to the European Convention for the Protection of Animals during International Transport was opened for signature. This protocol contains 7 articles, all of which are necessary to enable the EC to be a contracting party of the Convention, and it is understood that the EC through this procedure gains the status of a contracting party.

In fact, therefore, before the EC became a contracting party it had already established a Directive removing technical obstacles to trade in live animals by legislation.

There is pressure at EC levels to initiate action aimed at protecting animals against cruel handling during transportation.

(2)   EC COUNCIL OF MINISTERS DIRECTIVE ON PROTECTION OF ANIMALS DURING INTERNATIONAL TRANSPORT

The EC Council Directive (77/489) is dated 18th July 1977. The Directive as such consists of 9 articles containing definitions and the general obligations of the member states.

An annex is attached to the Directive, consisting of the provisions given by the European Convention.

In two chapters the regulations relevant to this meeting are given, comprising:

A.   Domestic solipeds and domestic animals of the bovine, ovine, caprine and porcine species

    I.   General provisions
       1. control by veterinarian
       2. construction of means of transportation
       3. accommodation of animals
       4. water, food
       5. loading, unloading
       6. welfare
       7. illness, injury

    II. Special provisions for transport by
       1. railway
       2. road

  3. sea
  4. air

B.   Domestic birds and domestic rabbits

A I 1   In the country of export the animals shall be
inspected by a veterinary officer. He is responsible for
3 things: 1) that the animals are fit for transportation;
2) loading arrangements; 3) a certificate stating identity,
that the animals are fit and records of the means of transportation. Unfit animals include: 1) animals likely to
give birth; 2) animals having done so during the preceding 48 hours. Veterinary officers in all countries
involved in animal transport are entitled to prescribe
rest and care for the animals.

A I 2   Adequate space and normal room to lie down should be
provided. Construction shall ensure protection against
inclement weather conditions and proper ventilation.
Boxes and cages should be labelled and easy to clean;
they must ensure the safety of the animals and allow
inspection. Floors must be solid, close-boarded and prevent slipping.

A I 3   There shall be segregation according to species, age
(young and adult should be kept separate), sex (uncast-
rated adults should be kept separate from females, adult
boars and stallions separate from each other).
Solipeds shall wear halters. Bovines shall not be tied
by the horns. Ties should be solid and allow for lying
down, eating and drinking.
Solipeds shall have individual stalls or their hind feet
unshoed.
Bulls above 18 months of age shall be tied and fitted
with nose rings for handling only.

A I 4   Water and appropriate food shall be given at suitable
intervals at least once per 24 h.

A I 5   Suitable equipment shall be used to prevent slipping and falling down.

A I 6   Consignments of livestock shall be accompanied by an attendant (unless carried in containers or other measures taken).
The attendant shall feed and water and milk, and cows are to be milked at intervals of not more than 12 h.

A I 7   Animals becoming ill or injured shall have veterinary attendance as soon as possible and if necessary shall be slaughtered without unnecessary suffering.

A II 1   The railway truck shall be marked, covered and constructed to prevent escape and ensure safety and adequate ventilation. Solipeds shall be tied facing one side or each other. Young animals may be loose. Large animals shall be placed so as to allow an attendant to move between them.
Violent jolting of trucks shall be avoided.

A II 2   Vehicles shall prevent escape, ensure safety and have roof and tying facilities. Partitions shall be rigid.
A ramp shall be carried.

A II 3   Animals shall not be on open deck unless in suitable containers which give protection against weather and sea. They shall be tied or in pens or containers. There shall be access to all animals. Accommodation and lighting facilities shall be available.
There shall be drainage.
The number of attendants shall be sufficient.
An approved instrument for killing shall be carried.
The vessel shall be provisioned with sufficient drinking water and appropriate foodstuffs.
Provisions for the isolation of ill or injured animals are required.
First aid treatment shall be given.

A II 4   Containers or stalls shall be used. Other arrangements for restraint can be permitted.
Extremely high or low temperatures and severe air pressure fluctuations shall be avoided.
An approved instrument for killing shall be carried.

B.   In addition to relevant articles of chapter II the following specific regulations are laid down:
Ill or injured animals shall be considered not fit for transportation. Those becoming ill or injured are to be given first aid and submitted to veterinary examination.
When animals are placed above each other in containers or on floors the falling of droppings on animals underneath shall be avoided.
Adequate food and water shall be available unless the journey is less than 12 h or in the case of chicks ≤ 72 h after hatching travelling ≤ 72 h.
Having been through a document like this one may of course raise several questions.
I have arranged my own questions in 4 groups:

1)   On what basis have these provisions been formulated?

2)   As they were formulated more than ten years ago are there better grounds today and so a need for reconsideration?

3)   Are there elements in the transporation of animals which are not covered by the document and therefore again a need for reconsideration.

4)   A final but possibly overwhelming question: how much welfare does the implementation of these provisions really establish for animals in transport?

THE BASIS FOR THE PROVISIONS

1.   Some are based on scientific evidence. For example, animals around parturition are already in a state of sometimes severe stress and that should not be increased. Adult boars, which may never have met each other, are known to be likely to fight, sometimes to the death.
Others of the provisions may seem to have no basis. For example, why is a cow in transport to be milked at intervals of

not more than 12 h when in some husbandry systems cows may be milked twice in 24 h with maybe 10 and 14 h intervals. There are no regulations outside the transportation situation.

I believe that some of the provisions are based on beliefs of the persons who wrote them and may very well be taken from national legislation which may have been conceived 50 years ago.
2.   Could it be done better today? Of course new knowledge is being provided but is there enough today to warrant reconsider- ation? I myself cannot answer the question but should like to put it in front of you. It may be better answered at the end of the meeting.
3.   Are there provisions which are lacking? I think this ques- tion might also be better considered at the end of the meeting.
4.   To what extent do such provisions ensure welfare? We can protect against many obviously bad things but transportation of live animals, under certain circumstances, is such that long dur- ation is bound to be a severe stressor and should be reduced when possible, for example, by avoiding long distance transportation of slaughter animals.

DISCUSSION

G. van Putten (*The Netherlands*) You mentioned that animals have to be fit for transport. Is this only relevant for animals which are expected to give birth, or are sick animals included? How wide is the definition of animals which are fit for transport?

H.C. Adler (*Denmark*) It means fit for transportation, and it is the respon- sibility of the veterinarian to inspect the animals in such a way that he can guarantee that the animals are fit for transportation. It is slightly differ- ent in the case of rabbits and birds, as a veterinary inspection is not requir- ed for those animals. It is therefore stated for those animals that diseased or injured animals are not fit for transportation.

G. van Putten I have asked myself whether pigs with leg or locomotion problems are really fit for transport. I doubt it. They have to be transported, of course, as they must be slaughtered, but you could argue that they are not fit for transport.

H.C. Adler They are not fit for transportation, and it should be possible to stop them being transported by means of the veterinarian's inspection. I should just like to mention our national legislation. Pigs such as you men- tioned cannot be transported to a slaughterhouse under Danish legislation. If they have broken legs, or other injuries, or particularly difficulties in moving up and down into the vehicle, they cannot be transported. The veterin- arian at the slaughterhouse is responsible, through his inspection of the live animals, to ensure that they have been transported according to the rules and regulations. If they have not been, he is obliged to report the matter to the police, and the man responsible for the transportation will be fined.

THE OFFICE OF INTERNATIONAL EPIZOOTICS (OIE) AND THE
INTERNATIONAL AIR TRANSPORT ASSOCIATION (IATA)

R. Moss
Regional Veterinary Officer,
Ministry of Agriculture, Fisheries and Food,
Hook Rise South, Tolworth, Surrey, UK.

I felt it important at the commencement of this seminar
to consider and emphasise the breadth of the basis of regulations
covering the transport of livestock. There are many thousands
of livestock moved every day for all sorts of purposes and by
all sorts of methods. There are also many thousands moved in a
particular way, and that is by air. Later in this seminar two
papers will be presented which deal with this particular mode
of transport.

During 1976 and 1977 a series of meetings was held between
officials within the Ministry of Agriculture in Great Britain
and veterinary officials from Australia and New Zealand, with a
view to initiating closer international co-operation in the
field of the air transport of animals.

Australia and New Zealand were particularly interested
because we were beginning to ship quite large numbers of breeding stock from Great Britain to their side of the world. How to
achieve international co-operation was at that time unclear, and
although initially it was thought that it could be achieved
through extension of the use of the International Air Transport
Association's manual of live animal regulations, there was very
little progress. I would just remind you that that manual is
produced by a small group of airlines which belong to the International Air Transport Association (IATA) and it is a very
definitive set of basic recommendations which are applied within
the IATA where the recommendations have the status of regulations.

In 1977 discussions took place with the Office of International Epizootics (OIE) in Paris about the possibility of that
organisation taking an interest in this subject. That was done
because that organisation covers some 99 nations, and it has,
in fact, for many years played a considerable part in the international movement of livestock in relation to both health and

welfare, and had already produced basic recommendations relating to such movements. It was considered that there was every possibility of adding to those recommendations further recommendations covering the transport of livestock by air.

In May 1978 a meeting of what became known as the Government/IATA Liaison Group (GILG) was held in Geneva. One or two people in this room were present at that meeting and it included people representing the Convention on International Trade in Endangered Species, the Zoological Society of London, the International Society for the Protection of Animals, the International Commission for Laboratory Animals, the United States Department of Agriculture and apologies were received from the World Wildlife Fund, the International Council for the Protection of Birds and from the Phillipines. At that meeting the United Kingdom said that, as presently constituted, this informal group had no official status and for that reason the UK could not participate fully in its activities. France, Germany and Belgium found themselves in similar positions, and it was decided that the International Office of Epizootics' initiative should continue. Approaches were now, therefore, formally made to the International Office of Epizootics and that organisation has set up a sub-committee of its Zoo-Sanitary Code Commission to consider the subject.

The sub-committee has met on two occasions and discussions have concerned stocking density in aircraft, the disinfection of aircraft, administrative procedures, problems relating to access to animals during flight, environmental control, tranquillisation of livestock prior to and during flight, conditioning prior to flight, penning systems facilities at airports and training of attendants. You can apply that list, of course, to all forms of transport. You will note that there are a great number of areas that need to be covered when we move an animal from point A to point B.

Professor Adler spoke of the European Convention, and he made the point that it was opened for signature on 13th December, 1968. That is 13 years ago. He spoke of the Community Directive 77/489. That is now four years old. Both contain references to air transport. There has been considerable attention paid

to air transportation, partly, I think, because it is still
novel, and also there are means of accompanying livestock and
therefore of monitoring livestock more easily than when they
are transported by road. You can also monitor movement by sea,
but as I have said air transport is still novel, easy to monitor
and carries large numbers of expensive animals.

This is the area in which the lessons learned have already
been applied. You will see from Dr. Watts' paper that in the
United Kingdom we are now able to make positive recommendations,
with a good scientific basis, on the stocking densities of
various types of animal. There are still questions that we
would like scientifically answered. Which way should a horse
face in an aircraft? What is the best form of ventilation?
What is the best form of penning? There are others, and these
I am sure will come up in the discussions. I simply wanted to
remind you that there is legislation, recommendations and regulations in this very specific area. I think the lessons that
have been learned there on the way in which that form of transport is monitored and the results of monitoring applied could
also be applicable to other forms of transport.

DISCUSSION

A. Hoogerbrugge *(The Netherlands)* I am involved with airline companies,
and one of my big problems is that all these airline companies are private
companies. Whereas we can discuss this subject here, the exchange of information between the airline companies is very limited. They train their em-
ployees well, but they do not exchange ideas or knowledge with the other
companies. How do we solve this problem?

R. Moss *(UK)* This, we hope, will develop through the association of the
OIE with the IATA even though that latter body has only a minority of the
airlines in the world as members. They have agreed to allow their expertise,
which they have coordinated through small groups of themselves, to be utilised through the OIE. For instance, they would be prepared to set up a
sub-group to look at the stocking density of a particular species and they
would then be prepared to have the findings of that group disseminated internationally through the OIE. We are beginning to tap the expertise of the
people who carry livestock by air.

A. Hoogerbrugge You still have the competition of the airlines who are not
members of this association. They have very low prices, very high densities
and then there are problems.

R. Moss The same problems exist with other kinds of transport.

P.J. O'Connor *(Ireland)* Is it the responsibility of transport companies to provide trained attendants or was it the intention of government agencies to have some course, or organise this training in some way?

R. Moss In the United Kingdom we have agreed to participate in the formulation of a programme of training if the airlines, and in particular IATA, wish to do so. I think that I am correct in saying that in Holland attendants travelling with livestock have to be proved competent by examination.

PHYSIOLOGICAL CHANGES INDUCED IN ANIMALS AT
LOADING, DURING AND AFTER TRANSPORT AND THEIR EFFECTS:
INCLUDING BEHAVIOURAL CHANGES

Chairman: E. *Wagner*

HANDLING OF SLAUGHTER PIGS PRIOR TO LOADING AND DURING LOADING ON A LORRY

G. van Putten

Institute for Animal Husbandry 'Schoonoord',
Zeist, The Netherlands.

INTRODUCTION

Transport of slaughter pigs has never been easy. It was difficult in old times, as old proverbs testify, and it is still difficult now, as is emphasised by an increased use of tranquillisers. However, tranquillisers neither solve the real problem which is that transport is obviously too great a stress for many pigs, nor add to the reputation of pork, because of the residues. Another disadvantage of tranquillisers is that they have a detrimental effect on the function of both the circulatory and the respiratory systems. However, it is exactly the performance of these two systems which is responsible for maintaining homeostasis of the animals' physiology.

Transport losses have become very low in most countries, often less than 1%. Professor Van Logtestijn will cover this aspect in his paper at this same seminar. However one thing should be very clear, transport death is by no means a merciful death, caused by heart failure. It is generally a painful process of suffocation, taking from ten minutes to two hours.

Apart from transport death there are other signs that slaughter pigs are often stressed beyond the limits of adaptation. According to Van Logtestijn (1970) up to 15% of all slaughter pigs suffer from being transported to such a degree that their physiology before slaughter and their meat quality afterwards are definitely abnormal. In these cases the meat quality is either PSE (pale, soft, exudative) or DFD (dark, firm, dry).

The breaking down of the biological integrity of 15% of the pigs involved due to being overstressed no doubt proves that the well-being (Van Putten, 1981) of all pigs in transport is seriously decreased.

PROBLEMS

Although generally transport itself is blamed for losses
and for a poorer meat quality, it is only part of the stress
to which pigs are submitted. When observing loading it is
quite obvious that some pigs are already in a pretty stressed
state on arrival at the transporting lorry. On the other hand,
lairage (and fighting) and stunning (restrainer or CO_2 tunnel)
can be very stressful too. In this paper we shall restrict
ourselves to the very first part of transportation, starting
with the removal from the fattening pen and stopping as soon
as the pigs are loaded on the lorry.

Two operations take place before the slaughter pigs are
actually on the lorry. The first phase is moving the pigs
from the fattening pen to the lorry. The second one is the
loading itself. The problems arising in these two stages are
quite different. It is therefore preferable to discuss them
separately.

Driving pigs from the fattening pen to the lorry

Fattening pigs are often kept on a fully slatted floor in
groups of about ten animals. The group remains intact until
slaughter, so pigs never meet strange animals after this first
day in the fattening pen. The pen is not subdivided into a
lying area and a dunging area. There is no substrate such as
straw for rooting and there is no bedding. Light is only
switched on during feeding, which means twice a day for half
an hour. The rest of the entire day and night they remain in
total darkness or in semidarkness, for there are no windows.
Room temperature is kept under control and set at a certain
level, all day and all night. Unpleasant and sometimes even
harmful gases from the stocked manure enter the pens. This
is what Kiley (1977) characterised as an 'oversimplified
environment'.

Dealing with stress has to be learned. Every individual
has to experience how to react to various stresses. After such
experiences events are less frightening and less stressful.
Our fattening pigs in their barren environment have hardly had

a chance to get acquainted with meeting other pigs, with taking
barricades, with light effects, with suddenly changing temperatures, with heavy exercise, with noise etc. etc. Nevertheless they are exposed to that kind of stress on their last day
of life. This is certainly not in accordance with the treaty
of the Council of Europe (1972), signed and ratified by all the
states represented in this meeting. The convention requires
all animals to be 'fit for transport'. The present situation
of pig fattening certainly does not meet this requirement.

It is still exceptional for the delivery of pigs to be
taken into account in the design of fattening houses. Sometimes
there is not even a door in the pens and thus the pigs have to
be lifted by hand over the partition. In more situations the
whole front of the pen, hanging on hinges above the trough, has
to be swung wide outwards to allow the pigs to be pushed over
the trough. In both cases not only are the animals stressed,
but also the men. Once in the feeding passage or in some other
corridor pigs have to be driven. If, however, the pigs in front
refuse to proceed, the pigs in the back of the group are vigor-
ously driven forward, often by means of electric prodders. The
entire operation functions fluently in some fattening houses,
but means hard labour in others.

Making pigs mount the loading bridge and enter the lorry

Unless the loading bridge can be used as a lift, it can
be very difficult to persuade slaughter pigs to ascend a loading bridge. These bridges generally make an angle of 30° with
the ground. Pigs simply refuse to climb the bridge. This is
not because of leg weakness, which increases the difficulties
for maybe 50% of the pigs (Penny et al., 1963; Prange, 1972).
Even if slaughter pigs are physically sufficiently fit to mount
the loading bridge, there seems to be a psychological barrier.

Fattening pigs, having no experience with such situations,
simply do not regard a loading bridge as a passable obstacle,
and they may even see it as a wall.

There is yet another problem. Once pigs are forced to
climb the loading bridge, they often hesitate to enter the lorry,
sometimes even running down the bridge again instead of entering.

AIMS

It would not be realistic to expect pig farmers to train their fattening pigs in order to overcome transport stress, although it is our experience that even a bit of exercise makes a large difference. If we cannot minimise the problems by training pigs to react properly, we just have to act the other way around and remove hindrances or at least diminish them. By doing this we have no less than four advantages.

1   The welfare of the pigs for which we are responsible is improved.

2   Loading is less laborious for the men involved.

3   Slaughter pigs have to endure less stress, which decreases the percentage of PSE meat.

4   There is little need for the use of tranquillisers. This saves money and adds to meat quality, because of the absence of residues. The residues of tranquillisers remain in the body for at least 24 h.

EXPERIMENTS AND EXPERIENCES

Tagging

In order to recognise pigs at the slaughterhouse they are usually given an eartag by the commission agent. In many countries the lorry driver is held responsible for the number of pigs he has loaded and for their welfare during transport. This is only possible if the animals can be traced. However, this procedure means that it is in the interest of the driver not to fix the eartags until immediately before transport and in his presence. In practice tagging is one of the duties of the driver.

Very often not all the pigs in one pen are ready for slaughter on the same day, so some of them are tagged and others are not. Thus tagging means a lot of chasing around and causes much commotion among the pigs, especially among those in that particular pen. It is much better to postpone ear tagging until the pigs are driven into a special delivery pen, from which they can directly climb the loading bridge. This delivery pen should be located at such a point where it is

easily accessible for the lorry, and where it can collect pigs
from several passageways. This special pen should preferably
be situated inside the building. If not, pigs might hesitate
to go into the cold air during winter. Being tagged in this
special pen is not nearly as stressful for a pig as being
tagged in its own pen. They seem to be overwhelmed by impres-
sions of all kinds once they have reached the delivery pen, and
hardly take notice of this extra pain. There is no need for
chasing in this pen, because such pens tend to be overcrowded
with pigs pressed firmly together.

By acting as has been described the pigs can be moved
quietly out of their fattening pen into some corridor. The
longer the pigs can be prevented from panicking, the more
easily the whole operation proceeds. Excited animals are
very difficult to handle.

If the lorry is equipped with a loading bridge operating
as an elevator and if this elevator is well fenced in, pigs can
be tagged on the elevator itself instead of in a special pen.
However, one needs this pen anyway to collect the animals which
are driven to the lorry.

Exit out of the fattening pen

Every pig pen, fattening pens included, should have a
well-functioning door, leading to the feeding passage or via
dung passages to some corridor. This is necessary for removing
sick animals, but also for delivery purposes. This door
should be situated in one of the corners of the pen and be
opened outwards. This allows one man to remove all the pigs
or only some of them, with the help of only a small wooden
screen. It is clear that there should be no obstacle such as
a trough in this opening. In that case farmers tend to use
electric prodders in the pen already. Sometimes the doors of
a pen have become so rusty, that they no longer function
properly. It even happens, that slaughter pigs are lifted out
of the pen by two men. This means hard labour and has a
detrimental effect on the pig, because the men do not put the
heavy pig back on the floor again, but just drop it outside
the pen.

The floor has to be stable

A pig is very reluctant to step on an unknown surface, which seems to be not quite stable. A wobbling plank, tile or segment with slats can be a cause of stagnation. Grids and unfamiliar slats have the same effect.

Keeping together

There is one thing a pig in distress always tries to maintain. This is contact with its penmates or else with any other pig. If possible they stay in touch with each other, but at least they remain in sight. This is one of the reasons why it is difficult to drive pigs around sharp corners. This also causes the problem in that pigs want to slip through the same small opening at the same moment, and get stuck.

We once constructed a simulator for analysing negative influences on fluent handling of pigs outside the pen (Van Putten and Elshof, 1978). We drove four inexperienced pigs from the same pen through a corridor with several obstacles which we wanted to compare. One pig had apparatus for radio-telemetric electrocardiography strapped on its back. The other three served as companions only. As long as these four animals could go on as a group and had ample room, nothing was wrong. However as soon as the passage narrowed (funnel-shape) so that only one pig at a time could pass through they reacted in such a way that all four tried to struggle through at the same moment and got stuck up and under each other. Their heartrate became as high as that of pigs climbing a low loading bridge (angle 30°, height 122 cm). Dividing the original corridor of 180 cm into three narrow passages of 60 cm each had a similar effect: all the pigs wanted to slip through the same opening at the same time, and again this obviously was as much a stress as a low loading bridge.

Doorways

Designers of farm buildings know that doorways have to be broad enough to let a feeding wagon pass. They do not realise, because nobody has ever told them, that doors should

not restrict the width of a passage at all. It should be possible to drive a group of pigs through a doorway without meeting
any hindrance or obstacle. Therefore doors should open outwards
or, if necessary, to both sides.

The influence of light

Pigs are careful in entering a strange room. If this
area is relatively dark, they hesitate. We observed 38 groups
of 4 pigs entering an unfamiliar bright room and the same pigs
entering an unknown relatively dark room (Van Putten and
Elshof, 1978). In about 50% of the observations the dark room
came first. Entering the dark room took 0.34 ($\pm$ 0.13) min,
entering the brightly lit room only 0.12 ($\pm$ 0.06) min. The
difference is statistically significant (P < 0.001).

In practice one can make good use of the result of this
test. As loading mostly takes place very early in the morning
while it is still dark, the influence of illuminating the
passage-ways and corridors can be great. Thus the light
switches should be within reach of the drover so as to enable
him to switch off nearly every lamp where the pigs are and at
the same time switch on all the lights at the point to which
the pigs are supposed to go. By repeating this procedure,
the drover (with his wooden screen) can direct the stream of
pigs easily. It is clear that sideways or possible escape
routes have to be closed beforehand with doors or with temporary fences. It will be equally clear that the delivery pen
also has to be brightly illuminated.

Loading itself

If a lifting platform is available this facilitates
loading and unloading no matter whether to and from upper loads
or to and from lower loads, both for men and animals. If not,
there is no reason to do it the hard way.

As pointed out before, it might well be true that pigs
without proper experience do not regard a loading bridge as
accessible. For the lower load it is easy to avoid a loading
bridge by situating the delivery pen on a platform at the same

height as the floor of the lorry. In that case moving pigs from this pen into the lower compartment of the lorry will cause no problem, provided there are extra lamps in the lorry and they are used properly.

However, if no loading platform is available and in some cases pigs have to be moved from a loading platform up onto the upper load, it is made easier by remembering another experiment which has been carried out.

The angle of the loading bridge

Again with radio-telemetric electrocardiography we measured (Van Putten and Elshof, 1978) the heart rate of one out of four inexperienced slaughter pigs. In this way we recorded the heart rate of 4 groups of about 20 pigs, ascending a loading bridge that led to a platform on a level of 122 cm above the floor (the height of the lower loading floor of a lorry). We used loading bridges with various lengths, making angles of 30° (as usual in practice), 25°, 20° and 15° with the floor.

TABLE 1

HEART RATE OF PIGS DIRECTLY AFTER CLIMBING A LOADING BRIDGE UP TO 122 cm. (THE ANGLES OF THE LOADING BRIDGES ARE DIFFERENT). EACH ANIMAL WAS ACCOMPANIED BY THREE OTHER PIGS (VAN PUTTEN AND ELSHOF, 1978).

Angle of loading bridge (deg.)	Number of animals	Heart rate (% of basic value)	Difference (progres- sive)
30	21	202 ± 23	25
25	20	177 ± 16	25**
20	19	160 ± 17	17*
15	20	139 ± 19	21**

* $p < 0.01$

**$p < 0.001$

The relation between heart rate and the angle of the loading bridge appears to be linear.

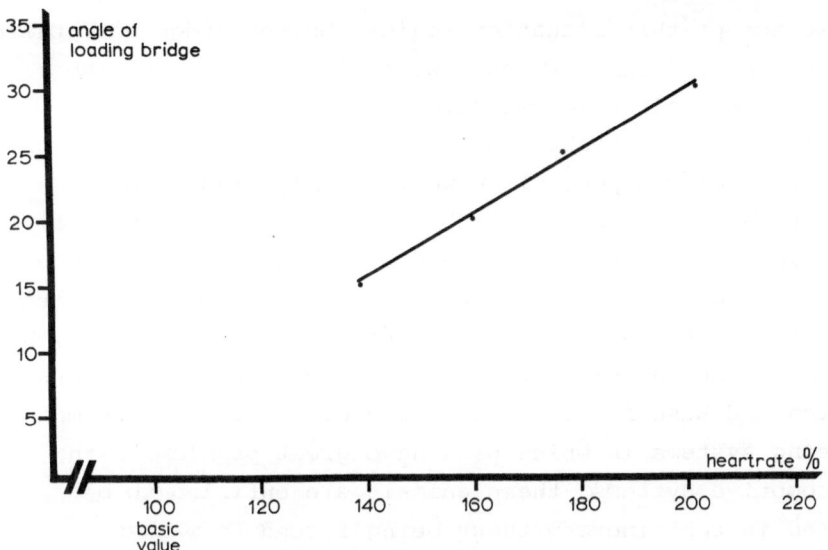

Fig. 1. The heart rate (as a percentage of the basic value) of pigs, directly after having climbed a loading bridge of 15°, 20°, 25° or 30° (Van Putten and Elshof, 1978).

The conclusion is obvious. If pigs have to climb into a lorry, a loading bridge should be provided which makes an angle of not more than 15° with the floor. This does make the bridge twice as long as the steep one with 30°. This long bridge (420 cm) can no longer be part of the transporting lorry, but should belong to the normal equipment of a pig fattening farm. Of course pigs should be prevented from falling down from this bridge by sidewalls. For reasons which have already been pointed out, this loading bridge should allow two or three pigs to mount together.

These provisions, combined with a long loading bridge, make use rather heavy. To keep this device mobile, it can be mounted on wheels, like aeroplane steps.

Leg weakness

If locomotion is painful, which may be the case according to the publications of Penny et al. (1963) and Prange (1972) in about 50% of all fattening pigs, all our care to get them on the lorry in good shape is more or less in vain. Every painful step is another stress for that particular animal. As

long as we accept this situation against better judgement, there
is only one way to transport such pigs without ignoring their
welfare. In that situation we have to enlarge the feeding
passages in order to be able to put a transport container
right in front of the pen. Then we load only pigs from the
same pen in one container in which there is room for about five
pigs, and leave them in this container during loading, transport,
lairage and internal transport in the slaughterhouse until they
arrive at the restrainer. Container transport is expensive, but
we have the technical skill to do it that way, as experiments
(Van Putten and Elshof, 1978) have proved. As long as we maintain housing systems in which pigs have great problems with
their locomotive systems, these animals are entitled to be
transported in containers without being forced to walk on
their hurting legs.

We should also make an exception for all pigs which
cannot stand on their hindquarters. Normal transport, as
described in this paper, does not apply to these animals.
Special transport is required, with very low carriers that can
be rolled right into the fattening pen.

Electric prodders

Electric prodders are handy. However, they are mostly
used on the wrong animals: those at the back, whilst those in
front have stopped moving. This paper would remain incomplete
without a warning to the effect that electric prodders do have
the same stressful effect on pigs as descending a loading
bridge or going through a funnel-shaped passage-way. Van Putten
and Elshof (1978) tested these effects. They found that
successive applications were even more harmful. The only
conclusion can be, that if at all avoidable, they should not
be used.

SUMMARY AND CONCLUSIONS

Transport does not start in the lorry. For pigs transport
starts in the fattening pen. By postponing eartagging until
pigs have arrived in a special delivery pen and by using a
wooden screen instead of an electric prodder, we can move pigs

quietly from their pen into the lorry. Of course there are
some necessary provisions:

- the fattening pen should have a well-functioning door,
leading to a passage-way;
- no obstacle such as a trough, for instance, should be
allowed in this doorway;
- passage-ways should have no narrowing parts, nor sharp
corners;
- floors should not have wobbling planks or other unstable
parts;
- light effects should be used by darkening the place
where the pigs are and lighting the area to which they
are to be moved;
- the loading bridge should not be too steep, and preferably not exceed an angle of 15° with the floor.

If the locomotive system of pigs is not affected, we
should be able to drive them onto the lorry without great
detrimental effects on their well-being.

REFERENCES

Council of Europe, 1972. European Convention for the Protection of Animals
        during International Transport. European Treaty. Series No. 65,
        Strasbourg.
Kiley, M., 1977. Behavioural Problems of Farm Animals, Oriell Press Ltd.
        Stocksfield.
Van Logtestijn, J.G., 1970. Veterinär-hygienische Aspekte der Stress-
        empfindlichkeit von Schlachtschweinen. Arch. Lebensmittelhyg.,
        3: 55-59.
Penny, R.H.C., Osborn, A.D. and Wright, A.J., 1963. The causes and incid-
        ences of lameness in store and adult pigs. Vet. Rec., 75, 1223-1240.
Prange, H., 1972. Gliedmassenerkrankungen bei Mastschweinen und der
        Einfluss unterschiedlicher Bodenausführungen. Monatsh. Vet.-Med.,
        27, 450-457.
Van Putten, G. and Elshof, W.J., 1978. Observations on the effect of
        transport on the well-being and lean quality of slaughter pigs.
        An. Reg. Stud., 1, 247-271.
Van Putten, G., 1981. Parameters for farm animal well-being, emphasizing
        the role of conflict-and vacuum behaviour. EAAP- Congress Zagreb,
        Comm. on Animal Management and Health.

DISCUSSION

G. von Mickwitz *(Federal Republic of Germany)* I would agree with you that pigs do not like to go into a dark room. They like to go towards the light. They never like to go towards the dark. That is very important.

G. van Putten *(The Netherlands)* They should of course not be blinded by too intense a light.

D.B. Stephens *(UK)* Dr. van Putten said that transport is sometimes painful, and I do not know quite what he means by that?

G. van Putten I mentioned leg weakness. Quite a number of pigs suffer from leg weakness and then walking, and therefore loading, is painful. The second time I mentioned it is when animals are under stress and are in such a condition that their meat will be PSE (Pale, Soft, Exudative) meat afterwards.

T.M. Leach *(UK)* You have made clear the snags connected with steep ramps which we ought to overcome. However, we are going to have steep ramps for some time to come. I wonder when you get pigs turning, on going up into the lorry, whether we might help this by having a narrow passageway so that the pigs go up in single file, following one another. Would this ease their entry?

G. van Putten It might help, but there should be no gap, or they will not go up at all. It is not so difficult to build a loading bridge on every large pig farm. A fattening house is expensive, and yet farmers may not think of building a loading bridge.

D. Lister *(UK)* I think we can all agree with the general recommendations given by Dr. van Putten on the ways of handling animals during slaughter, but it is very difficult to demonstrate that there are any substantial benefits from doing these kinds of things except in terms of those animals which are going to die or are going to produce PSE meat. How do you begin to demonstrate the benefits of using humane handling procedures?

G. van Putten That is an easy question to answer. I had to be very brief in my presentation, and I am glad that you have asked about this. It is much less laborious for the men if the stock is easy to handle. If you have to load pigs at 4 a.m. it is a bad job anyway, and if you have to push them around or lift them it makes the job even more unpleasant. It is easy to persuade the farmer of this and to make suitable arrangements.

D. Lister But then we are thinking of the welfare of the farm workers and not of the animals.

G. van Putten Yes, but your question was how to persuade farmers to do these things if I understood you correctly.

D. Lister I should like to see how you demonstrate the benefits of humane handling procedures. I agree that there are benefits to the farmer, but what about benefits to the pigs?

G. van Putten You have less PSE meat afterwards, but PSE meat is not such a problem for technologists, I am told. Then it is very difficult to persuade people of the benefits of looking after the welfare. I think it is not so much a benefit as a responsibility. It may be difficult, and I know that it is difficult to appeal to the responsibility of people, especially farmers, who are not aware of the condition of the pigs when they arrive at the slaughterhouse.

R. Dantzer *(France)* Firstly, I did agree with your description of the effects of tranquillisers. I would not argue about the fact that tranquillisers have to be used for transporting pigs but I agree that the decrease of cardiovascular and respiratory functions help the pig to withstand shock. This is mainly due to the cardiovascular changes.

My question concerns motion sickness. I have transported several hundred pigs and I have only noticed vomiting pigs two or three times. Do you know whether pigs are really sensitive to motion sickness?

G. van Putten They are, and especially so if they have been fed just before the journey.

R. Dantzer Yes, but there is another question which is then raised, and that is whether vomiting is a good criterion for the assessment of motion sickness in pigs? Do you know of any work on the systematic investigation of pigs' sensitivity to motion sickness using centrifugation, for example, or some other means? This is very important in terms of animal welfare.

G. van Putten I have seen pigs vomiting during transport in the lorry, especially when they had been fed, but I do not know of anybody who recorded motion sickness in any other ways.

TRANSPORT OF BROILERS

A.R. Gerrits[1] and K. de Koning[2]

[1] Spelderholt Institute for Poultry Research, Beekbergen,
[2] Institute of Agricultural Engineering, Wageningen,
Netherlands.

INTRODUCTION

In the EC every year about 2 000 000 000 (2 x 10^9) broilers are transported from the growers to the processing plants. Transport distances vary considerably. Even in the Netherlands broilers are often transported more than a 100 km. Several types of chicken crates or containers are used within the EC.

Since 1977 there has existed within the EC a general regulation that protects animals from being mishandled during transportation within the Community. Based on this EC regulation, a new regulation (Anon. a, 1980) was created specifically for the transportation of farm animals within the Netherlands. In this regulation, for example, the minimum space per broiler required, the need for adequate ventilation etc., are described.

The catching, loading and hauling of broilers is meeting with more and more problems. One problem is that the cost of catching is gradually rising. Catching broilers is unpleasant work, which often has to be done at night in dimly lit surroundings; therefore, it is increasingly difficult to employ people for this kind of work. Since it is such unpleasant work which has to be done quickly, the broilers are not always gently handled. Furthermore, the manner of catching and further handling and the transport units are not optimal; the opening in the top of the widely used plastic (and wooden) crates in which the broilers are transported is too small, not promoting optimal treatment of the animals. In practice the broilers are squeezed through the small opening in bunches of four or five. Little imagination is needed to realise that this filling procedure is unpleasant for the animals. A comparable situation is found in the containers where the broilers are loaded through an opening in the side.

The rough handling of broilers also has an adverse effect on the dressed product. Carrying the broiler by one leg involves the risk that leg or hip joints may become damaged and this can lead to internal bleeding. Rough handling may also cause broken bones which places the broiler in a lower quality grade (B-quality). Haematomas in leg and breast muscles occur frequently when broilers are not handled with proper care. In particular, the dropping of crates (stacks of 8) from a great height can cause haematomas in the breast meat close to the keel of the breastbone. Rough handling of the birds may induce stress and adversely affect the quality of the final product. Published data indicate that stress may reduce the tenderness of the meat. Looking at these factors there appear to be sufficient grounds for critically investigating the whole procedure of catching, loading and handling of broilers.

Efforts to improve the entire bird transport system have been going on for years. The demand for more efficient systems has become more urgent in the past few years due to rapidly rising costs, the difficulty in finding good workers and the growing demands of animal welfare organisations.

The Institute of Agricultural Engineering (IMAG) and the Spelderholt Institute for Poultry Research (IPS) have been asked to study and improve the entire catching, loading and transporting process in the Netherlands. The Fund for Poultry Interests and the Fund for the Welfare of Domestic Animals have promised financial support for this investigation.

CATCHING AND LOADING SYSTEMS WIDELY EMPLOYED IN THE NETHERLANDS

In the loading method most generally employed in the Netherlands 16 crates, placed on pallets, are transported between truck and broilerhouse with the aid of a tractor fork-lift. In the broilerhouse four catchers fill the crates, whilst two people bring the empty crates and remove the full ones. On the truck the full crates are stacked by two people, whilst a third man places 16 empty crates on an empty pallet. The catching of the broilers in the broilerhouse is usually done by picking up a bunch of four or five animals by the legs and squeezing them through the opening of the crate.

Another method which is not as frequently practised but nevertheless merits attention, involves catchers crawling around on the floor and picking up two animals by the sides with both hands. By this method the wings remain closed; the risk of injuries is reduced, especially because the broilers are not picked up by the legs.

The crates may also be transported from the broilerhouse to the truck on small pallets or a roller conveyor but the first mentioned method is cheap and reasonably rapid.

Hand-loading also occurs. In the case of broilerhouses with doors in the side-wall this is practicable. Two people pick up the animals in bunches of five and hand them over to the 'carriers' who walk to the truck with a bunch in each hand. The truck is parked by the door and two or three people there stack the crates on the loading floor of the truck.

FURTHER MECHANISATION OF CATCHING AND LOADING

In the Netherlands, and elsewhere, attempts are being made to mechanise catching and loading still further, to improve working conditions, and to increase the capacity and the quality of the process (Jewett, 1959). A recent Dutch system is the Boersmatic, in which mats are laid in the broilerhouse several hours before loading. The animals settle down on the mats and at loading time the mat is slowly pulled to one end of the house and rolled up in very dim light. The animals fall onto a conveyor belt which conveys them to a 'crating machine'. Crating takes place at high capacity; a maximum of one crate is filled each 5 seconds (Anon. b, 1979). This crating device is part of the Kähler loading system from Germany, in which six catchers place the animals on a long conveyor belt running the length of the broilerhouse (Kähler). Both methods (Boersmatic and Kähler) use a similar system of supplying the empty crates and removing the full crates. A truck with empty crates has first to be unloaded before it can be loaded with full crates.

A totally different system has originated in Israel (Anon. c, 1980). The broilers are scooped up by a very manoeuvrable mini-tractor fitted with front-loading prongs (60 - 80 per scoop) and dumped into a container placed on a platform.

From this container four to six people pack the broilers into crates. In other systems in the United States and Italy the birds are transported loose to the truck by means of conveyor belts or by air via pipes. The various tiers of the truck can be loaded from a platform whose height is adjustable (Anon. d, 1970). The trucks may be provided with a structure of cages with doors or flaps on the outside or with floors provided with conveyor belts. Catching in the house is still done by hand, but both the conveyor and the hopper to the pipe can be moved easily, so that walking distances are minimal. One advantage of these systems is that no crates enter the processing plant.

Important disadvantages of these cage and/or belt systems on trucks are their high price and the fact that the broilers have to be unloaded directly onto the feed conveyor leading to the hanging line, which means that no buffer supply can be formed.

In England, Sun Valley has developed a container which is filled in the house and loaded onto the truck with a forklift (Farrant, 1977). A disadvantage is that the containers are too big to be placed under the hanging line, as crates usually are, and have to be placed behind the hangers so that the hangers have to take each broiler from the container behind them. In this way hanging is less convenient than when crates are used. In the United States a similar container is tipped to unload so that all the broilers fall onto the feed conveyor of the hanging line (Shackelford et al., 1976).

Again in England, Anglia Autoflow and D.B. Marshall have recently developed a large module containing 12 plastic trays. This module is filled with broilers in the house and then transported to the processing plant. There the trays are automatically removed from the module and transported to the 'hanging crew'. It is easy to remove the broilers from these trays.

Only a small number of processing plants are equipped with a mechanical de-stacker for the broiler crates. In many cases the full crates are dropped onto a roller conveyor and so the top crate from a stack falls down about two metres.

During manual de-stacking the crates are not always kept
in a horizontal position so that sometimes all broilers are
pressed together along one side of that crate. Removing the
broilers from the crates and hanging them onto the shackle of
the overhead conveyor is done rather roughly in practice. In
a processing plant which slaughters 6 000 broilers per hour, a
man from the 'hanging crew' has to hang one bird every 5 sec
onto a shackle.

Some processors are thinking of conveying the birds on a
belt without crates, since this may be expected to result in an
increase in the hangers' efficiency, as they can then pick up
the broilers more easily. With a special type of crate which
has a lid forming half of the upperside, the broilers can be
dumped onto this feed conveyor.

Although this method enables transportation to the hanging line to be mechanised to a great extent, the tipping of the
crates to empty them (as in the aforementioned systems) is not
advisable due to the risk of injuring the broilers. Neither is
it desirable that live animals should be treated in such a way.

REQUIREMENTS TO BE MET BY A NEW SYSTEM

The exploratory phase of the investigation (literature
study; talks with growers, slaughterers, advisory officers and
other experts; personal observations) has yielded many aspects
which have to be considered in improving the transportation of
broilers. For example:

- The whole process, 'from house to shackle', should be
  considered. Grower, hauling contractor and slaughterer
  should be involved.
- The size of the transport unit (crate, cage, container)
  should be such that it cannot be thrown around. This
  implies that transport units have to be moved by mechanical means.
- The opening of the transport unit should be large and in
  such a position that both filling and emptying can be
  carried out as efficiently as possible and that the broilers cannot be mishandled.

- The transport units should be filled as close as possible to the place where the broilers are, and unloaded at the hanging line, so that handling of the animals is reduced to a minimum.
- The transport system should meet the minimum legal requirements for the transport of live poultry (height in the crate 0.24 m or 9.5 in; space/kg broiler 0.018 m^2 or 28 in^2).
- Ventilation should be adequate.
- The transport units should be easy to clean.

A new concept is being developed based on these requirements.

THE SPELDERHOLT-IMAG CONCEPT

Large containers (1.85 m x 0.85 m x 0.25 m or 73 in x 33 in x 9.8 in) which can be stacked one on top of the other, so that the one above serves as a lid for the one below, constitute the transport units (Figure 1). The containers are brought into the houses with a forklift and filled there with broilers and stacked five high. With the Boersmatic system or another mechanised supply system the broilers can be picked up by hand, using a tractor scoop. A stack of full containers is brought outside with a forklift. The next stack of five is placed on top of it, making a stack of ten containers which is put onto the truck at one time. A simple lid is placed on the top container to prevent the broilers from getting out of the container.

A truck with a loading floor of 7.5 m x 2.5 m can take 80 of these containers, twice four stacks of 10 containers, which are placed with their long sides in the direction in which the truck moves.

A wide space is left between the rows (Figure 2) for ventilation. In the processing plant the stacks of containers are removed from the truck with a forklift and placed on a roller conveyor which transports them to the de-stacker. The containers are then moved to the hangers by means of another roller conveyor. This last stretch of roller conveyor may have to be covered to prevent broilers from jumping out of the

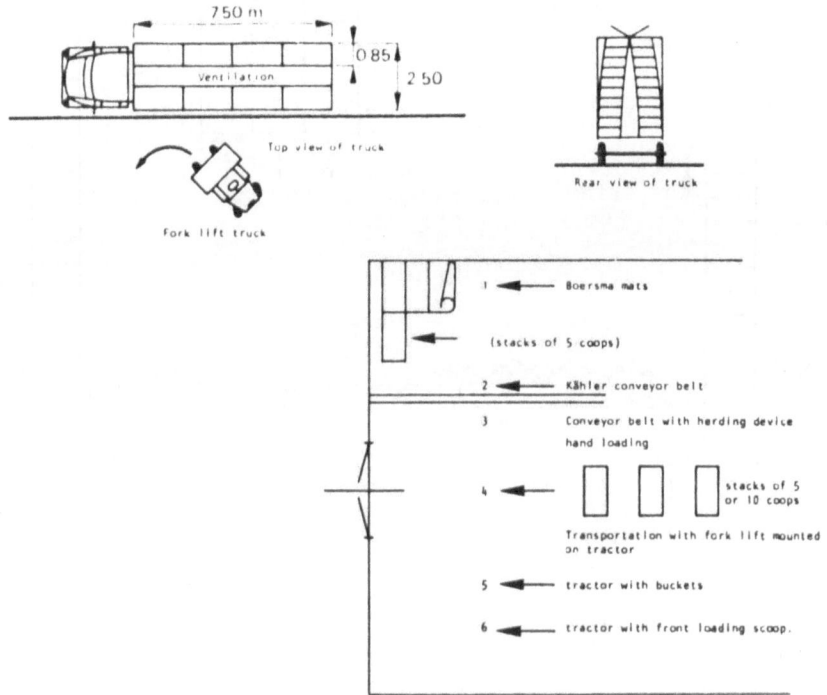

Fig. 1.   Loading of broilers at the broilerhouse with large open top scoops.
On each truck 80 scoops containing 5 000 broilers.

containers.   The hangers can now easily pick up the broilers
and place them onto shackles.   Investigators in the United States
found a similar open system to generate a higher hanging capacity
per person and fewer injuries of the animals.

The empty containers are cleaned in a washing machine and
then stacked mechanically in stacks of ten.   Both the full and
empty containers can, if necessary, be placed in a buffer or
storage area.   In hot weather the full containers can be placed
against a ventilation wall to provide the broilers with fresh air.

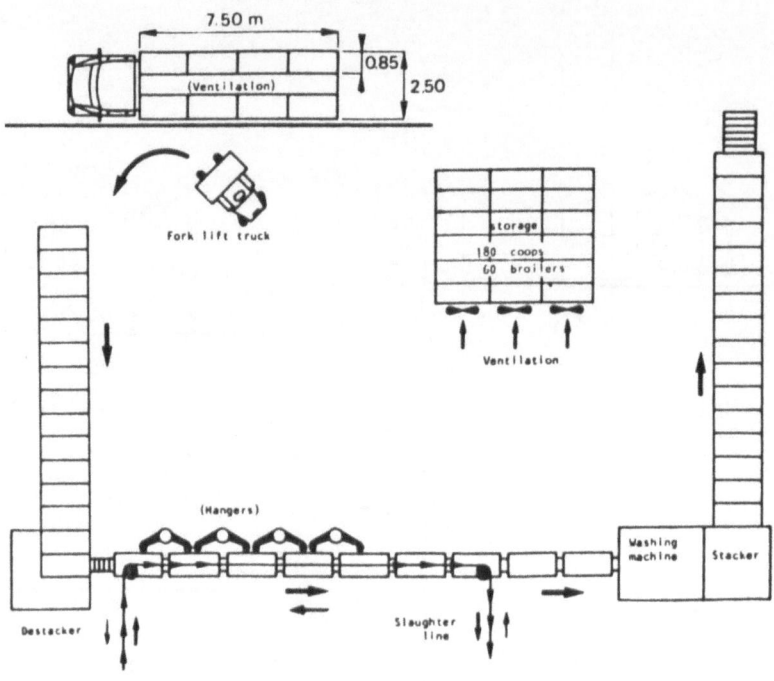

Fig. 2. Layout of reception centre at a slaughterhouse when broilers are
transported in large open scoops.

The first experiments with the above described containers
have already been carried out under practical conditions. The
system proved to be satisfactory. The birds, once in the containers, were very calm and only a few tried to jump out. In
addition no problems were encountered at the processing plant.
Since these preliminary studies have proven to be successful,
plans are being made to use this system on a commercial scale.

REFERENCES

Anon. a, 1980. Beschikking wegvervoer vee en pluimvee. December 1980.

Anon. b, 1979. Oplossing voor het laadprobleem van mestkuikens in zicht
Pluimveehouderij, 30, 12-13.

Anon. c, 1980. Broiler Scoop. Poultry International, April, 54.

Anon. d, 1970. Bird suction system halves loading job. Broiler Industry
33, 46-52.

Farrant, J., 1979. Modules halve catching costs. Poultry World, May, 20-21.

Jewett, L.J., 1959. Handling and processing broilers in Maine. Part 1.
Costs and efficiencies in assembling live broilers for processing.
Part 2. Quality losses in live broilers and methods of handling to
reduce bruising and improve efficiency (Thesis Univ. Maine).

Kähler, F. Verfahren und Vorrichtung zum Abpacken von Lebendgeflügel
insbesondere Mastgeflügel (Beratung Kähler, Norderheistedt, BRD).

Shackelford, A.D., Holladay, H.H. and Thomson, J.E., 1976. Live birds
automation reduces labour 30%. Broiler Industry, 39, 28-30.

DISCUSSION

__M. Jespersen__ _(Denmark)_ There can be problems of tenderness of the meat if
the living animals are badly treated. Do you have more than one reference
for this?

__A.R. Gerrits__ _(The Netherlands)_ I have only one reference and that says that
stress can induce less tender meat.

__M. Jespersen__ It has been proved, though?

__A.R. Gerrits__ Yes, in the United States.

__M. Jespersen__ We saw these open crates of broilers. Have you had any problems
with specific strains of broilers not staying in the coop before they are
hung?

__A.R. Gerrits__ No, but I think it may be a real problem. I think that it will
not be so easy to use these open crates in some parts of Europe and in the
warmer regions of the world because birds are very flighty. In Holland, the
birds are rather calm. I know there is one strain which is not so calm, but
I hope we can transport them using the same system. I know that this is a
risk in the system.

__M. Jespersen__ But the rest of the strains have been tried and there have been
no problems?

__A.R. Gerrits__ They have been very, very calm.

__P.V. Tarrant__ _(Ireland)_ Dr. Gerrits, can you give any estimate of the amount
of stress related problems in these birds, such as transport deaths, injuries
or meat quality defects? Have you any quantitative estimate of the size of
this problem?

__A.R. Gerrits__ It is not so easy to give an amount. It differs with the processors and the systems used. The people from one company told us that after
using this open coop system, they had almost 100% A quality. With previous
systems they had had 40% A quality, and now they have 80% A quality. This
means that this system has improved meat quality, and also animal welfare,
in this situation.

__P.V. Tarrant__ Dr. van Putten mentioned a figure of about one third of transport
deaths in pigs. Is the figure for broilers greater or lesser than that?

A.R. Gerrits I have no data on that. It depends on the slaughterhouse and on other factors.

M. Jespersen I am going to present some figures in my paper concerning damages to broilers.

M.E.T. Watts *(UK)* The design of your crates would appear to prevent ventilation whilst the crates are on the vehicle in that your design shows ventilation holes on the long side, which becomes the inside when the crates are stacked on the vehicle.

Secondly, we in England have very variable climatic conditions. How would you reconcile your system with protection from rain, sunshine and frost and still get adequate ventilation?

A.R. Gerrits In this system we have a large ventilation area between the two rows of stacks. We can bring forced ventilation into that space; that is merely a technical question. In the event of rain or snow, it is possible to cover the crates with canvas as is done already. We have not yet worked this out but there are ways of preventing the birds from being mishandled in bad weather.

INJURIES DURING CATCHING AND TRANSPORTATION OF BROILERS

M. Jespersen

Danish Research Institute for Poultry Processing,
Smørkildevej 8, DK 3400 Hillerød, Denmark.

ABSTRACT

The purpose of this work is to indicate, under Danish conditions, the amount of injuries received by broilers during catching and transportation. The investigation began at the farm with the catching of the animals and continued afterwards at the processing plant.

The injuries were divided into five different types. It was found that the parts of the birds that were most subjected to damage were the wings, the backs, and the drumsticks. It was observed that the most critical steps of the catching phase were apparently the transfer of the birds to the crates, the handling of the filled crates at the processing plant, and the hanging of the birds onto the shackles.

INTRODUCTION

These studies were made in connection with plans for making a more mechanised, possibly better, catching system. The processing plants in Denmark have become interested in a more mechanised system because of the very inconvenient working hours and poor working conditions. An altered system would be even more attractive if in addition it resulted in less damage to the birds.

There are very few references in scientific publications concerning injuries to broilers. An English article (Jee, 1980) mentions a loss of £1 million per 40 million broilers due to injuries. This loss is partly due to downgraded birds and partly to removal of injured limbs. In addition the number of down-graded birds is estimated to be 9 - 12% of the total production.

This value is in agreement with the one found by the processing plant where the present studies were made. They found that 12% of the production over a six month period were second class birds.

It is difficult to tell the loss of profit as the price difference between first class and second class broilers fluctuates from month to month. This study indicated that the total number of injured birds was far greater than the number of second class birds. Accordingly, many first class broilers are found with some kind of injury.

At present there are practically no studies that determine whether or not there is a quality difference between flawless and slightly injured broilers. Possibly the consumer is accustomed to a product which ought to be classified somewhere between first and second class.

MATERIALS AND METHODS

The catching and transportation phases are defined here as the treatment of the bird from the time that it is living freely on the farm until it is hung on the shackle before the stunner.

In this investigation a batch of broilers was studied throughout the catching and transportation phases, thereafter the injuries were tabulated by random sampling after the birds were first slaughtered and plucked. The counting is thus made considerably easier.

It must be taken into consideration that additional injuries occur before the counting is done due to the very hard treatment in the plucker. After consultation the veterinarians and after sampling at the conveyor, it was discovered that the injuries occurring in the plucker could be distinguished from the earlier ones. The earlier injuries were not darkened by blood accumulations.

As it was not possible to describe every injury individually, they were divided into groups. Most of the injuries could be included in five major types:
1   dark spots on the drumsticks
2   dark spots on the breast
3   dark spots on the back
4   extravasation on the wings
5   fractured wings

It was decided that dark spots larger than 20 - 25 mm in diameter should be included as injuries.

RESULTS AND DISCUSSION

It is apparent that the treatment of the birds can be divided into four separate phases:

1   catching of the birds;
2   transferring to the crates;
3   unloading of the crates at the processing plant;
4   hanging of the birds onto the shackles.

1   The observed catching of the broilers was relatively gentle. This was most likely due to the skill and good teamwork of the workers concerned. The workers agreed that this phase could be much more brutal if a new member joined the team.

2   At this stage the injuries began to occur. The following procedure was used: the broilers were caught, ten at a time, carried to a truck, and then handed to a man on the truck. The workers had devised a procedure that let ten broilers pass through the opening of the crate at a time. It was found that this step caused damage to flapping wings, since the broilers must pass through a relatively small opening with a certain speed.
The observed catching crew received hourly wages in contrast to many other workers who are paid per broiler. It is therefore very likely that the observed treatment was more gentle than normally seen. It could be suggested that the workers let five broilers instead of ten pass into the crate at a time, although this requires greater physical exertion.

3   It should be possible to unload the crates without further damage to the birds. This was not found to be the case. Due to the physical strain on the workers, the upper crates were very often allowed to fall freely to the roller conveyor in front of the hanging station which is often a drop of approximately two metres. It is not difficult to imagine the dark spots that can occur on the breasts and backs of the birds that have been through this treatment. This step is usually done manually

although a few processing plants have automatic de-stackers.
which allow a more gentle treatment.

4   To begin with the birds were removed from the crates.
The speed with which this was done did not make it possible to take special care of each bird, thereby allowing
further damage to occur to flapping wings as they hit the
edges of the small opening of the crate. It was observed
that the workers of the hanging station were often
unnecessarily brutal towards the broilers. For instance,
the broilers received an unreasonably hard pull as they
were fixed into the shackles, a pull which can cause
strain to the thigh muscles.

Everywhere in Denmark stunning is carried out using
electric voltage. Unfortunately, it is found that many processing plants use too high a voltage, thereby causing muscle con-
traction, as found in the breast for example, to be so violent
that the muscle attachment breaks, causing bruises to form. It
is debatable whether or not this has any effect on the welfare
of the bird, but in any case it gives a poorer quality of the
final product.

As seen in Table 1, it was first of all the wings, the
backs, and the drumsticks that were subjected to injury. As
previously mentioned the procedures causing damage were
primarily transferring of the birds to the crates, unloading
of the crates at the processing plants, and hanging of the
broilers.

CONCLUSION

This has been a very small investigation that gives a
little information about the total injuries that are found
today on broilers in Denmark. Before making any possible
changes in the catching and transportation systems, additional
studies should be made in order to determine how many injuries
each operation inflicts.

It seems at present already apparent that changes should
be made in both the design and the handling of the crates which
may be done with minimal investment.

TABLE 1

FREQUENCY OF DIFFERENT DAMAGES FOUND ON BROILERS

	spots on the drumsticks	spots on the breast	spots on the back	extravasation on the wings	fractured wings	
		Types of damage				
Average (%)	4.0	2.7	5.7	7.8	1.6	Date: 1981, June 2
Highest	5.2	1.7	–	11.1	1.1	Total batch 26 000 Age, days 39
Lowest value (%)	3.2	5.2	–	2.6	1.9	Average living weight 1 360 g
Number of measurements	7	7	2	7	5	Dead birds found in the crates 59 Time of transportation 2½h
Average (%)	5.1	1.8	4.0	3.8	0.6	Date: 1981, June 3
Highest	–	–	–	4.8	1.0	Total batch 27 000 Age, days 40
Lowest value (%)	–	–	–	2.7	0.1	Average living weight 1 380 g
Number of measurements	2	2	1	3	3	Dead birds found in the crates 24 Time of transportation 2 h
Average of total (%)	4.2	2.8	5.1	6.6	1.3	
Standard deviation (%)	0.78	1.2	1.6	3.3	0.6	

REFERENCE

Jee, D., 1980. Live bird handling can be streamlined. Poultry Industry, Oct. 1980, p 32-33.

DISCUSSION

E. Wagner (Luxembourg) Are the injuries you mention really all caused by transport or handling, or are some present already prior to removal from the house?

M. Jespersen (Denmark) There may be a little damage, but I find it hard to imagine that animals living freely in a house should incur damage from that source.

E. Kallweit (Federal Republic of Germany) I wondered about your mention of the high voltage used for stunning because with pigs, voltage is reduced to avoid damage and to improve meat quality. I wonder what difference there is in the stunning of poultry and pigs, and what voltage you use to stun poultry.

M. Jespersen Normally they use a voltage in the region of 70 - 80 volts. I think it is much higher for pigs. Unfortunately here they used about 150 volts, and that is much too high.

W. Sybesma *(The Netherlands)* For how long is the voltage applied?

M. Jespersen The speed of the chain you saw was 4 500 per hour, so I think the length of the stunners was approximately 1.2 m.

W. Sybesma That would give a time of about 10 seconds.

M. Jespersen I do not think they would be given electricity for 10 seconds. That is much too long.

T.M. Leach *(UK)* I am interested in the point about high voltage stunning, because it was shown some years ago that high voltage stunning of pigs will give fractures if the pigs are on the floor but not if they are in a conveyor. Poultry are not restrained and I find it very difficult to understand how this type of damage occurs due to stunning.

M. Jespersen It is very easy really. If you watch broilers being stunned you will see that they give a very big flap of the wings, so we very often see dark spots inside here in the pattern of the wings.

T.M. Leach To move on to a different subject, do you think that these voltages are killing them or not?

M. Jespersen Well, I think that 150 volts is likely to kill the birds.

VARIOUS TRANSPORT CONDITIONS AND THEIR INFLUENCE
ON PHYSIOLOGICAL REACTIONS

G. von Mickwitz

Freie Universität Berlin, Fachbereich Veterinärmedizin,
Klinik für Klauentierkrankheiten und Fortpflanzungskunde,
Königsweg 65, 1 Berlin 37, Federal Republic of Germany.

It is very difficult for the modern slaughter pig to compensate for the strain caused by transport.

One reason for this is its genetically conditioned latent myopathy. Many pigs, therefore, suffer after transport from a disturbed muscle metabolism. The consequence of this disturbed muscle metabolism is a rapid decrease in muscle pH after slaughter (pH 45 under 5.6), a poor water binding capacity and pale meat (PSE).(Figure 1).

PSE meat not only has characteristic visual indications, it also has (and this must be clearly stated) bad palatability. Enquiries of slaughterhouse staff regarding the frequency of PSE will often obtain the reply, "PSE is not a big problem for us".

Here it must be particularly pointed out that normal meat inspection does not record the real frequency of PSE (Tables 1 and 2). In our investigations we found that at routine meat inspection of 237 058 slaughtered pigs only 99 were classified as PSE (0.041%), but when we judged the carcases by the pH 45 minutes *post mortem* at the same slaughterhouse the frequency was 239 PSE in 1 160 examined pigs, i.e. 20.6%. We define PSE as a clinical problem from three aspects, viz, genetic, transport and handling following transportation,including the method of stunning. Over the past ten years we have been looking at transport conditions in relation to clinical reactions and in relation to the mortality rate (Figure 2). In this respect we have looked at the reaction of the heart rate during normal transport conditions as it has been influenced by the times of feeding prior to commencement of transportation. We have also considered the reaction of the circulatory system and the filling of the auricular arteries and veins (Figure 3). Study of these last mentioned reactions tells us something about the circulatory function within the animal as it has been affected through transportation.

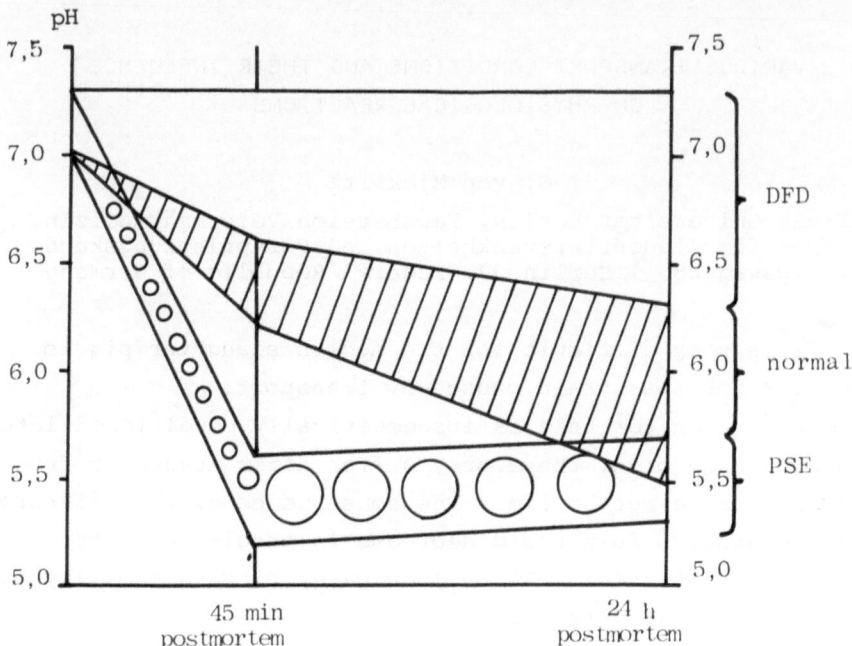

Fig. 1. Decrease of pH in the muscle of pigs after slaughtering.

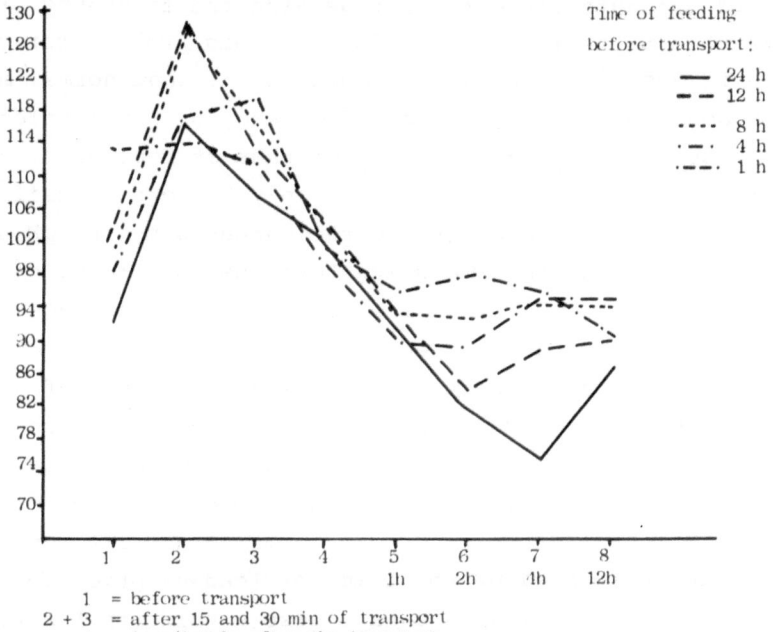

1 = before transport
2 + 3 = after 15 and 30 min of transport
4, = immediately after the transport
5 - 8 = rest period

Fig. 2. Change of heart frequency of pigs during transport on lorries influenced by the time of feeding before transport.

TABLE 1

OCCURRENCE OF PSE PIGS JUDGED BY 'NORMAL' MEAT INSPECTION METHODS*

Year	No. of slaughtered pigs	No. of PSE	PSE (%)
1976	253 263	47	0.018
1977	287 614	49	0.017
1978	310 047	48	0.015
1979	237 085	99	0.041

TABLE 2

OCCURRENCE OF PSE PIGS ACCORDING TO THE pH_{45} LEVEL (ALL CARCASES WITH A pH LEVEL BELOW 5.3 45 MIN AFTER SLAUGHTERING WERE REGARDED AS PSE POSITIVE)*

Year	Number of carcases	Number of PSE	PSE (%)
1979	1 160	239	20.6

* Both investigations were performed in the same slaughterhouse.

In other investigations we examined the influence of the loading density on the body temperature, heart rate and respiratory rate (Figure 4). There is a strong correlation between loading density, heart rate, respiration rate and body temperature, especially if transportation occurs during high temperature periods. Loading density will influence meat quality.

During hot weather conditions all pigs arrive at the slaughterhouse with a high body temperature. After one hour, however, all pigs will return to normal values for heart rate, respiration rate and body temperature, with one exception: all transports with a high loading capacity (density less than 0.4 m^2/100 kg KGW). These pigs arrive at the slaughterhouse with an extremely high body temperature and the pigs will retain high body temperature for more than one hour.

Pigs slaughtered with such a high body temperature will have a higher percentage of PSE meat than pigs slaughtered with a normal temperature. The limit for the loading density for slaughter pigs is 0.5 m^2/100 kg KGW (Figure 5).

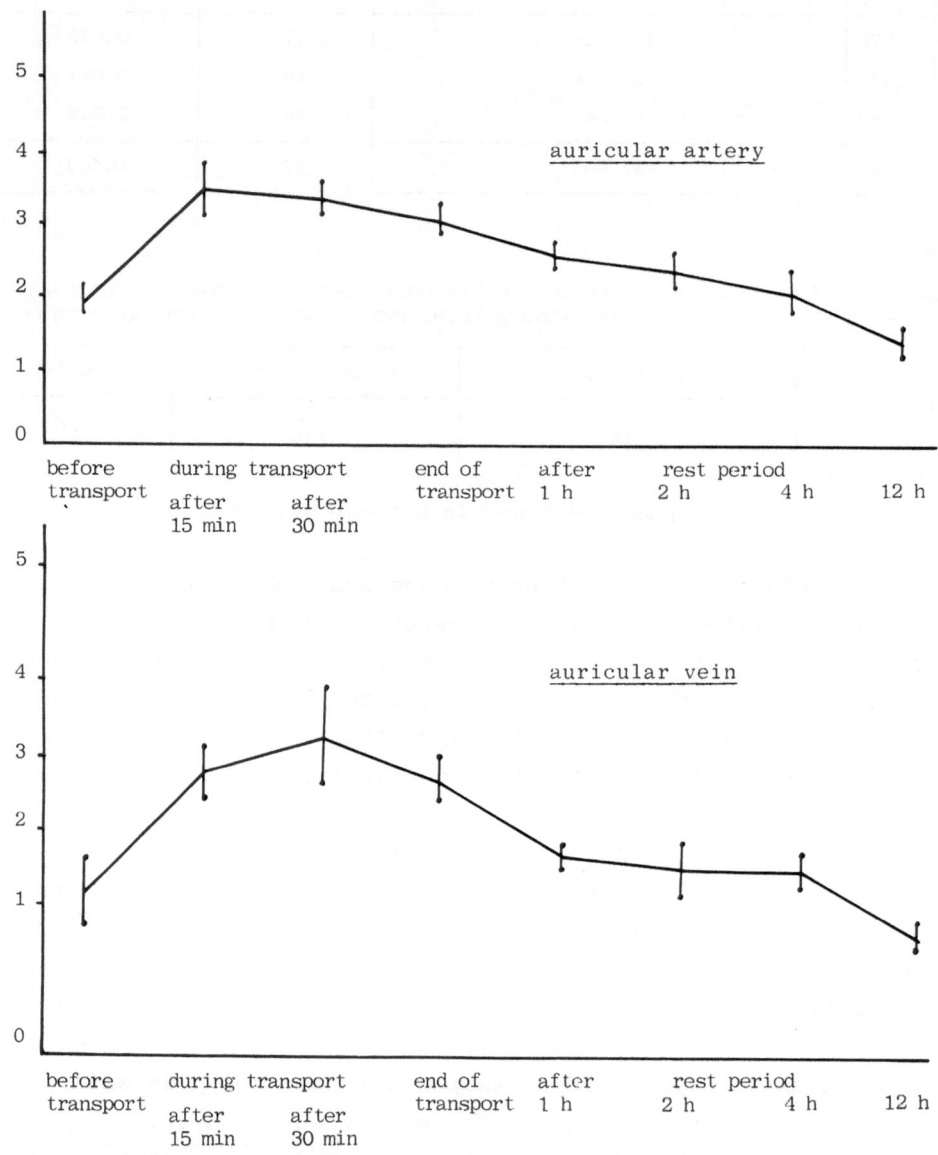

Fig. 3. Filling of the auricular artery and vein before, during and after transport.

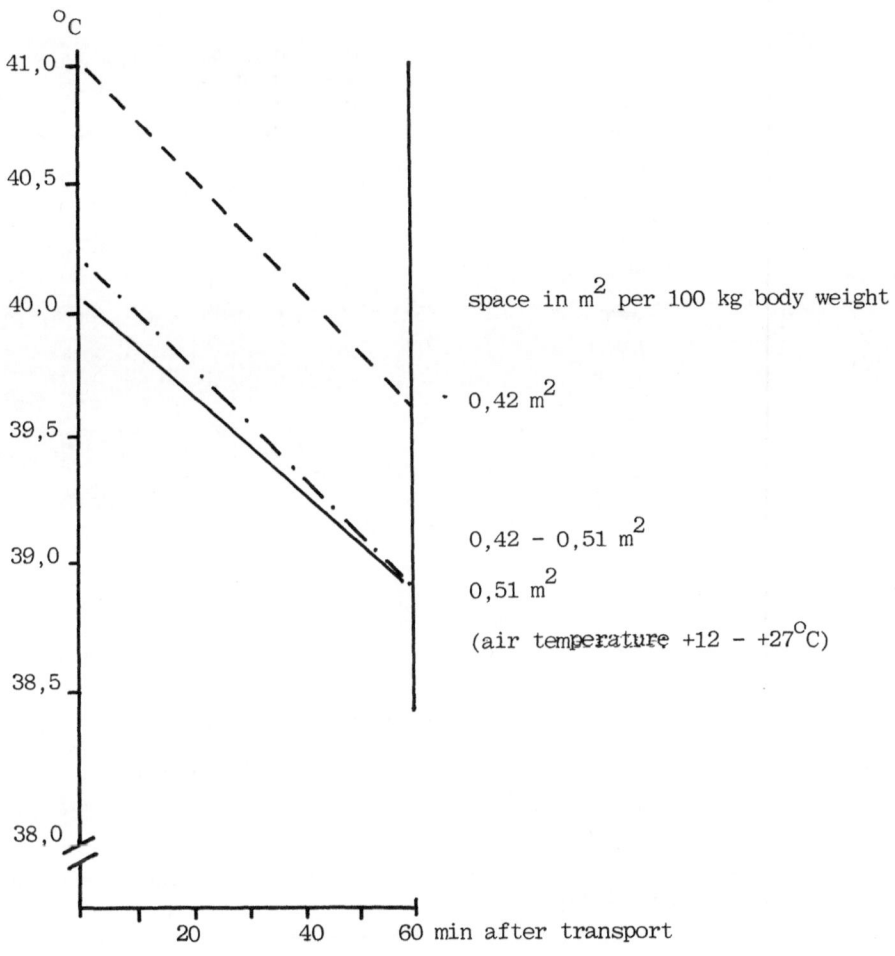

Fig. 4.   Body temperature of pigs related to loading density during
transport; the temperature was taken immediately after transport
and after a rest period of one hour.

During the summer when there are high air temperatures
and there is high humidity we have to allow 15% more space.
There is a strong correlation between the mortality rate during
transport and the so-called 'Schwülefaktor', (air temperature
plus 2 x water content in mg/l) (Figure 6).

50

temperature of the
day of transport C°

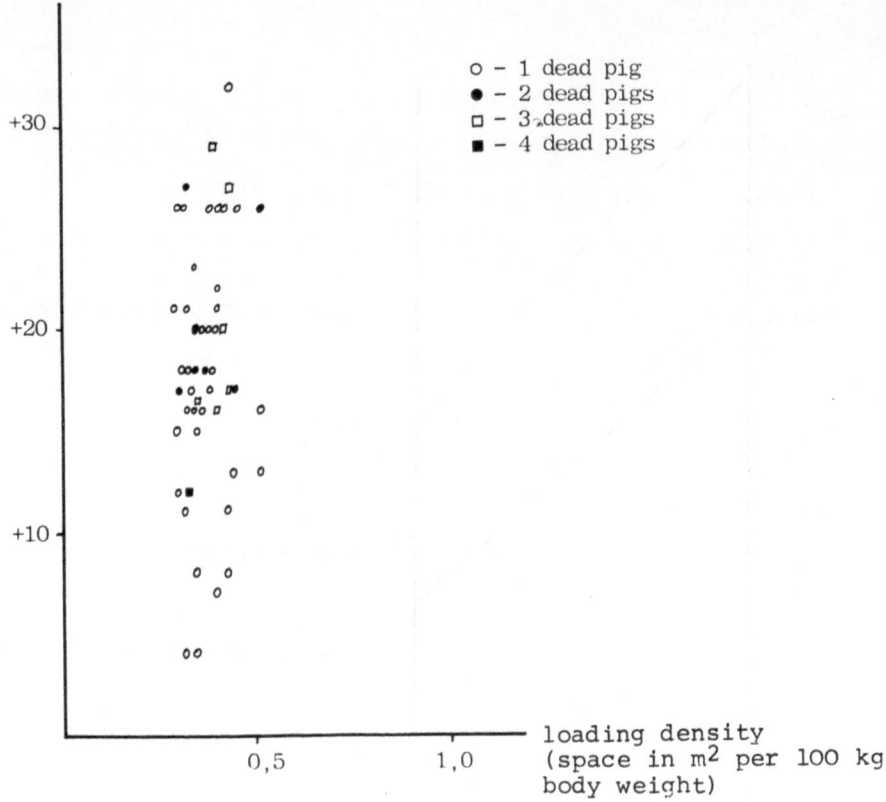

Fig. 5. Transport mortality in relation to weather conditions (temperature and loading density).

The size of the ventilation openings is very important in this connection (Figure 7).

Extremely high body temperatures following transportation, as a physiological reaction, is a bad sign indicating that the transport conditions have been bad.

In one slaughterhouse we found that the number of animals dying during transportation was higher during the second part of the week than was the case on Mondays and Tuesdays (Tables 3 and 4). We found that on the days with higher pig mortality, larger lorries had been used, each with a capacity exceeding 30 pigs and

51

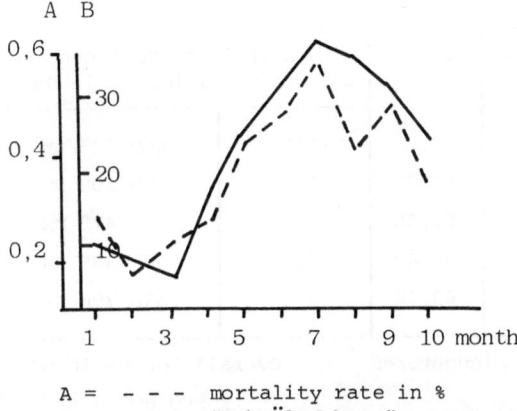

A = – – – mortality rate in %
B = – – – – "Schwülefaktor"

Fig. 6. Average mortality rate during transport from January to October in relation to the 'Schwülefaktor' (humidity in mg x 2 + air temperature in C°).

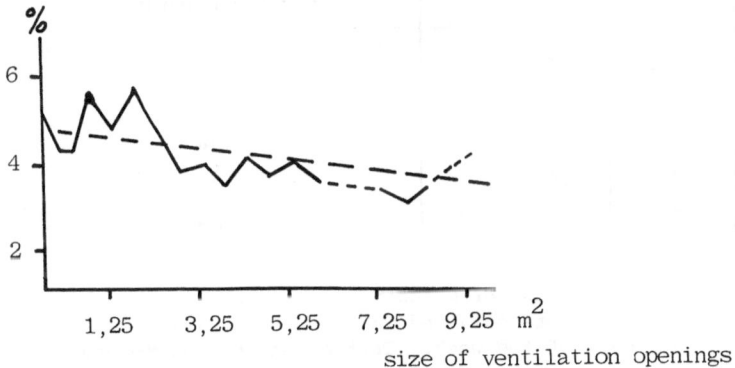

size of ventilation openings

Fig. 7. Size of ventilation openings (m^2) on the lorry in relation to the mortality rate during transport.

drivers of those lorries did not separate the pigs into groups of 10 to 15 animals. In this way the slaughterhouse lost more than 200 000 DM. On Monday and Tuesday more pigs were transported in smaller lorries. The mortality was lower but there was also a difference between these two days of 0.41% to 0.48%, which made a profit difference of nearly 80 000 DM (Figure 8).

TABLE 3

DAY OF THE WEEK AND PERCENTAGE OF LOSSES DURING TRANSPORT

Day of the week	Mortality during transport (%)		Va-Ko	Slaughter days (No.) considered and financial losses in DM	
Monday	0.41	0.27	65.6	307	409 346 DM
Tuesday	0.43	0.31	65.27	311	479 298 DM
Wednesday	0.60	0.37	62.55	321	598 475 DM
Thursday	0.57	0.33	58.49	320	567 385 DM
Friday	0.54	0.34	63.54	331	536 796 DM

(2.6 = 100%) Number of animals slaughtered   Overall losses in DM

1 663 047   2 590 800

Losses during transport 8 636.0

TABLE 4

PERCENTAGE OF ANIMALS TRANSPORTED IN LORRIES WITH A LOADING CAPACITY OF MORE THAN 30 PIGS, RELATED TO THE DIFFERENT DAYS OF THE WEEK

Day of the week			Number of weeks considered for the experiment
Monday	79.5	2.41	14
Tuesday	80.5	2.066	14
Wednesday	90.86	1.875	14
Thursday	87.29	2.091	14
Friday	82.14	2.107	14

(The considerably lower figures for Mondays can be explained by the fact that on Sundays the animals are only fed once in the morning, and not twice a day as on the other days of the week. That means that on Mondays the animals had fasted at least 24 hours before transportation).

We found that the reason for this was that all pigs slaughtered on Monday had a significantly smaller weight of stomach contents than the pigs transported on Tuesday (Figure 9). The explanation for this was that on Sundays, pigs in large pig farms are only fed once a day and thus pigs transported on Mondays have effectively fasted for 24 h prior to transportation. To prove the influence of the time of last feeding prior to transportation on the weight of the stomach content, we fed pigs 3 kg feed, 1,4,8,

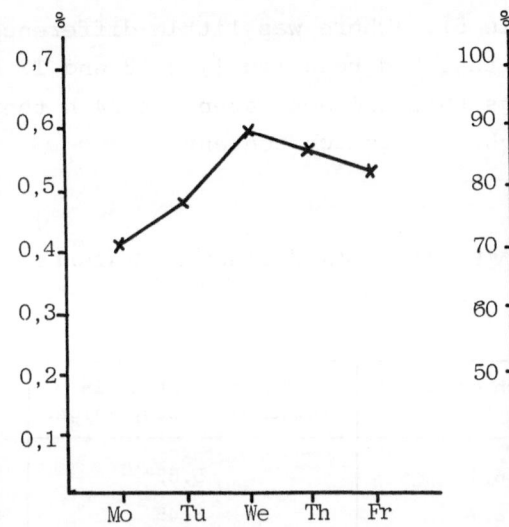

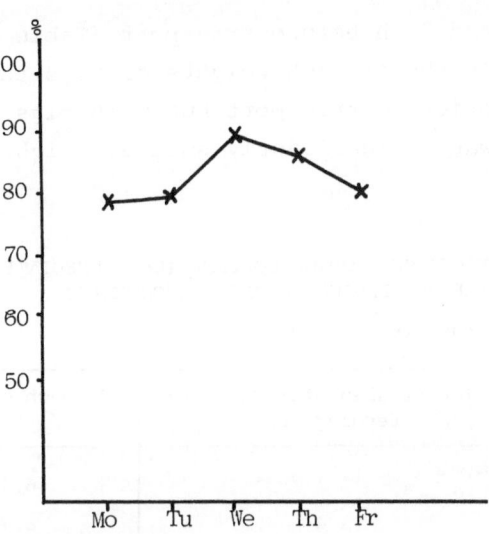

Fig. 8. Number of pigs (in %) that died during transport.

Number of pigs (in %) transported in lorries with a loading capacity of more than 30 pigs.

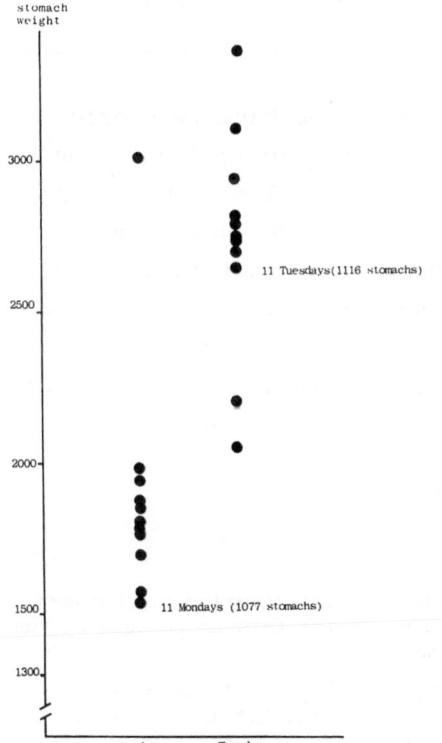

Fig. 9. Average stomach weight of slaughtered pigs, slaughtered on 11 different Mondays and Tuesdays.

54

and 24 h before transport (Table 5). There was little difference
in the stomach weights of pigs that had been fed 1, 4, 8 and 12 h
prior to transport but with pigs that had not eaten for 24 h there
was a signiicantly smaller weight of stomach content.

TABLE 5

WEIGHT OF STOMACH CONTENT (kg) IMMEDIATELY AFTER SLAUGHTERING IN RELATION TO
TIME OF FEEDING BEFORE SLAUGHTERING

Amount fed: 3 kg

Time of slaughtering after feeding (h)	Stomach content	Stomach content in relation to body weight
1	6.4	5.62
4	6.7	5.45
8	5.1	3.53
12	5.3	3.84
24	2.2	1.44

If animals die during transportation we have to consider
this as the worst result of an extreme physiological reaction.
Of course, there are many reasons why an animal succumbs during
transportation and each factor is worth considering. Some factors
alone would not kill a pig but, coupled with other factors, might
well do so. The same thing must be said for all the factors which
are able to produce bad meat quality. Very often bad meat quality
is the result of the sum of many factors arising from poor transport and handling conditions on the way from the farm and to the
stunning box.

DISCUSSION

R. Dantzer (France) What is your explanation of the relationship between
the loading density and mortality rate? Do you think this is related to the
problem of air circulation between the animals?

G. von Mickwitz (Federal Republic of Germany) I think there is a strong
correlation with the size of the ventilation openings, but the other point
is that you must allow 0.45 or 0.5 m^2 per 100 kg, and you must add an extra
15% space in hot weather conditions. We researched this in over 2 million
slaughter pigs.

There is also a strong correlation with air conditions and humidity.
If you plot a graph of this you get almost the same curve over the year for

humidity and mortality. I think Allen in England obtained the same figures, the only difference being that his maximum space was 0.2 m^2 and ours in Germany was 0.6 m^2.

We made about 600 animal transports. We allowed them 0.5 m^2. All the dead pigs were those which had less than 0.4 m^2 per 100 kg. This is very important. If you have more than this you will get dead animals.

You have a high temperature and high respiration rate. All these pigs will have a PSE meat. All the people who will work with the meat after death, not only the slaughterers, know this.

M.E.T. Watts *(UK)* I would like to support Dr. Mickwitz. British observations on aircraft transportation tally exactly with the figures that he is giving on densities and environment in his paper.

G. von Mickwitz Something can be done if you are looking for good transportation. These are the figures for almost twenty years. We started with less than 0.3% dead animals and it remained like that until 1971, when we stopped looking at the figures, and continued in another slaughterhouse. When we looked for good transportation we found that it was possible to reduce the mortality from 0.4% to 0.3%.

A.T. Fensvig *(Denmark)* I would like to know if you are counting only mortality during transport or including mortality in the lairage when you talk about mortality.

G. von Mickwitz I mean those which are dead on arrival at the slaughterhouse.

E. Wagner *(Luxembourg)* I know that hot weather is very closely connected to PSE frequency, but has there been any investigation made into the effects of very cold weather?

G. von Mickwitz That is a very difficult question. The physiological reactions are quite different in cold weather. I cannot answer that. We were only investigating the effects of hot weather conditions.

There are so many factors. People will now take fewer animals in hot weather and more in winter so that you could have the same situation. You can sometimes find very high body temperatures in winter, but that is due to transport conditions and not to alarm.

R. Dantzer As this seminar is about the welfare of transported farm animals, I have a very general question in relation to meat quality. Can the incidence of meat quality variation be used as a way of assessing poor transport conditions? This is a fundamental question, but it is probably too early to ask it, perhaps it would be better kept for the final discussion.

G. von Mickwitz You cannot say that if an animal arrives with a high temperature, heart rate and so on after bad transport conditions, its meat will all be low quality, but you can say that a higher percentage of its meat will be of a low quality. I have been working in animal welfare for twenty years and I believe it is better to show this aspect rather than simply to say that these are bad conditions and must be improved. From my point of view this way is better.

A. Cuthbertson *(UK)* I would like to raise one question which, with so many experts here, might be answered later. I believe there is evidence that the longer the delay between last feed and slaughter, the greater the risk of loss of carcase weight when you compare animals of similar liveweight. This suggests that there may be some loss of water from the carcase. Is there any evidence of this? Has anybody done any studies on composition of tissues from animals which have a delay between last feed to slaughter? Do they agree with the suggestion that there may be a water loss related to the stress problem we have been discussing?

W. Sybesma *(The Netherlands)* There is a study in Holland, still going on, which says that the quality changed in that the water holding capacity changed as did the water content.

G. von Mickwitz In which case, when you feed them or when you do not feed them?

W. Sybesma When you feed them. The shorter the period the more water you lose so that it is not worth feeding the animals shortly before slaughter as this gives a greater weight loss.

G. von Mickwitz We were surprised to find nearly the same content after 12 h as after 8 h or 4 h. If you want to influence stomach weight, you should give the animals water during that time. You should stop feeding for 24 h after transport. There are many other questions along the same lines. What about testing after transport? That is the same. You can lose 2 DM/pig if you give them resting time.

D. Lister *(UK)* Pigs go into the post absorptive phase 16 h after the last meal, and this is the time when they start to lose carcase tissue, usually in the form of glycogen. This is tied up with water and this is where the water loss comes from.

P.V. Tarrant *(Ireland)* We looked at this problem in relation to sugar-feeding of pigs and we found that when we took two groups of pigs of similar liveweight and held one overnight with access to water only and gave the other group access to sugar solution those which had had water only showed a very significant loss when carcase weight was taken as a percentage of live-weight. I think there are some other studies to support that. Carbohydrate alone appears to be sufficient to offset the loss of weight during overnight lairage.

PHYSIOLOGY OF THE TRANSPORT OF CATTLE

T.M. Leach

Agricultural Research Council, Meat Research Institute,
Langford, Bristol, UK.

ABSTRACT

   Transport inevitably puts cattle in a stressful situation and considerable attention has been given to changes in various biochemical and physiological systems as indicators of the development of stress reactions during and after transportation.

   Exposure to stress can impair the immune response, facilitating multiplication to pathogenic levels of viruses and bacteria normally present as commensals. The well recognised entity of shipping fever is widely accepted to result from stress causing lowered vitality, which leads to viral infection and this facilitates bacterial invasion.

   Exhaustion as a result of transportation reduces the animal's physiological capacity to respond to vehicle movements and so predisposes to injury, particularly bruising.

   The commonest physiological effect of transport is loss of body weight, which is largely, but not exclusively, attributable to withholding food and water. A direct correlation exists between weight loss and duration of journey. There is some conflict in the experimental evidence on the proportions of loss attributable to voiding gut contents and to tissue wastage.

   Depletion of glycogen reserves in the carcase musculature as a result of psychological or physical stress, particularly if food and water are withheld, can affect development of rigor mortis. Less than normal production of lactate is then responsible for a high final pH and the phenomenon of dark cutting beef.

   Amelioration of all these factors requires attention to be given to the training of stockmen in the handling of animals in transit, to improving the design of loading/unloading facilities and vehicles, and, not least, to the behavioural requirements of the stock.

INTRODUCTION

   Although transport of livestock is often an economic necessity, from the animals' point of view it is an unnatural activity that inevitably exposes them to a variety of hazards. Stress may arise from loading and unloading, mixing with strange stock, withholding food or water, physical injury or climatic conditions. In this context, the term stress is used to imply prolonged and/or intense stimulation capable of causing pathological changes, the state termed 'overstress' by Ewbank (1973).

Individual response can be affected by breed or sex, by care in handling, by design of equipment, including consideration of behavioural needs, and by duration of the journey.

Hails (1978) has reviewed transport stress in animals. His comprehensive treatment embraced a number of conditions not normally considered under the physiological term stress, e.g. mastitis, but this approach ensured that the diverse effects of transport were adequately covered. The present paper is principally concerned with subsequent publications relating to the transport of cattle, although, where appropriate, reference is also made to earlier papers.

EFFECT OF TRANSPORT ON BIOCHEMICAL AND PHYSIOLOGICAL SYSTEMS

In an attempt to develop an index of stress, Pérez-Fernández and Willoughby (1976) transported Holstein-Friesian heifers, aged 1.5 - 2.5 years, in a lorry for 1½ h and Holstein and Ayrshire cows, aged 2.5 - 4.5 years, for 2 h. In samples taken 4 h after the journey, both cortisol levels and total white cell counts showed a marked increase over pre-transport values. The white cell count had returned to normal values within 10 h, while the cortisol level was normal within 48 h.

Levantine et al. (1977) examined blood samples from 350 calves and steers, aged 13 - 14 months, transported 35 km for slaughter. Loading caused sharp increases in cortisol, glucose, lipids, protein and urea; the steers were more affected than the calves. In animals then fasted 24 h before slaughter glycogen values were halved and 48 h fasting reduced them seven fold, compared with animals killed on arrival. Lactate greatly increased during handling and transport but fell during 48 h fasting.

A comparison of various blood serum values in weaned and transported calves was reported by Crookshank et al. (1979). Sera from calves that were weaned and immediately transported for 12 h were compared with those from calves that were weaned only and also from calves weaned 30 days earlier. Transport produced a definite increase in cortisol, but weaning alone resulted in only a small increase. Normal levels returned within two days of weaning and four days of weaning and transport. Increases in cortisol levels decreased as calves became accustomed to handling. Slight increases occurred in creatine phosphokinase (CPK), lactic dehydrogenase (LDH), glutamic oxaloacetic transaminase (GOT) and glutamic pyruvic transaminase (GPT) after weaning, but these were not exacerbated by transport. No differences were observed between values from weaned and weaned and transported calves in levels of alkaline phosphatase, cholesterol, creatine, glucose, urea nitrogen, uric acid, total protein, Ca, Cu, Fe, Mg, inorganic P, K and Zn.

Simensen et al. (1980) observed 17 day old Norwegian Red bull calves transported by lorry 5 - 30 km to a market and 320 km, an 8 h journey, on the following day. In addition to the significant, but transitory, increases in serum cortisol levels and white cell counts reported by earlier authors, they noted significant, but brief suppression of serum IgG values and suggested that changes in the levels of circulating immunoglobulins could be used as a quantitative measure of the stress of transportation.

The effect of transport stress on the levels of serum enzymes was also studied by Groth and Gränzer (1975). Three groups of calves weighing about 65 - 75 kg were transported 223 - 265 km by lorry and six weeks later, when their body weights were approximately 107 - 165 kg, they were transported 266 - 310 km. Serum enzyme activity rose significantly in some animals after the first journey and in all after the second journey; GOT increased by 7 - 38%, GPT 29 - 83%, LDH 14 - 42% and CPK 30 - 308%. The rate and degree of rise was greatest for CPK, but it also fell most rapidly. Sick calves showed markedly greater increases than those caused by transport. Groth and Gränzer (1977) compared conventionally managed calves with early weaned calves. At the start of the trial, when three weeks old, both groups were transported 164 km and five weeks later they were transported 275 km. Total leukocyte count, GOT, CPK, LDH, cholesterol, lactic acid and free fatty acids (FFA) all rose in both groups. Early weaned calves had the larger CPK and FFA increases, but hyperglycaemia was higher in the conventional group. Growth inhibition was less after the second transport than after the first. The authors

postulated that the smaller carbohydrate reserves of the early
weaned calves forced them to increase CPK production in the
muscles to utilise FFA for energy production.

Deutschmann et al. (1976) transported 35 cattle, aged
between 1½ and 20 years, 15 - 49 km in temperatures varying
between 4° - 32^OC. Serum values of GOT, LDH and aldosterone
(ALD) all rose after transport and continued to increase to the
time of slaughter, 20 - 50 h later. The rise in ALD was most
marked and was related to transport distance (over 20 km), age
and ambient temperature.

Samples taken from 10 Friesian bulls after transport and
sale in a market were found by Kriesten et al. (1976) to have
significantly reduced values for total protein, inorganic P,
Na, chloride and Ca, but K was not significantly lower. Total
lipids, glucose and Ca/inorganic P ratio were all significantly
raised, but creatine was not significantly affected. Kriestin
et al. (1978) investigated changes in serum lipids and fatty
acids in bulls weighing 517 ± 78 kg stressed by transport and
passage through a market. They sampled the animals before
despatch and three days later. There were significant increases
in total lipids, cholesterol and cholestryl esters. Levels of
nearly all fatty acids were raised and the correlation patterns
between lipid fractions and fatty acids, and between individual
fatty acids, were altered.

Histological changes in the musculature of bulls slaughtered after transport were examined by Velinov and Enev (1978).
Six animals were moved 25 km and 3 were then slaughtered
immediately, while the other 3 were laired for 24 h before
slaughter. Examination of *M. longissimus dorsi* samples taken 2,
24, 48, 72 and 144 h after slaughter showed no differences
between the two groups in muscle structure or *post mortem* changes.
It was concluded that immediate slaughter after a short journey
was satisfactory.

Dotta et al. (1976) examined blood samples from 162
cattle, comprising either one month old calves or animals
2 - 4 years old, after long journeys at various times of year
by different means of transport. In the great majority pH was
normal or slightly lowered, pCO_2 values were consistently

decreased. Chloride levels were always raised and some groups showed a slight hypocalcaemia. There was no evidence of other significant changes in electrolytes. In a study of 18 - 30 months old Limousin cattle transported by road or rail for over 500 km Mouthon et al. (1976) found that blood values of both pCO_2 and pO_2 had decreased, while pH increased. They suggested that oxygen deficiency could be a determining factor in the occurrence of degenerative myopathies in these circumstances.

Damron et al. (1979) examined changes in the rumen ciliate protozoa of calves stressed by weaning and transport. The first group received hay at the auction barn and order buyer barn, the second received a high energy ration with antibiotic at both barns, the third had been weaned 30 days before despatch and received a 50% concentrate ration *ad libitum* throughout, the fourth received the high energy ration with 50 g $NaHCO_3$/kg body weight at both barns. Seven species of ciliates were counted. In all four treatment groups organisms/ml rumen fluid fell from the level at the farm of origin, through the auction barn to the buyer barn. Slight recovery in numbers had occurred in groups 2 and 4 on departure from the latter.

The effects of different transit temperatures and of fasting were studied by Fischer and Augustini (1980) in Simmental bulls fasted for 22 h before transport. Groups of cattle were carried for 4 h in a lorry held at $2^{o}C$, $12^{o}C$ and $36^{o}C$. Heart rates were unaffected, but the rectal temperatures had risen in the groups at $36^{o}C$ by the end of the journey. Samples from the *M. semimembranosus, M. semitendinosus, M. longissimus dorsi, M. psoas major* and *M. biceps femoris* taken at slaughter on arrival showed that the cattle carried at $36^{o}C$ tended to have the lowest glycogen and highest lactate and glucose levels. Adenosine triphosphate (ATP) and creatine phosphate values were largely unaffected, nor was the final pH elevated. Groups transported similarly were then fasted 48 h before slaughter. Muscle glycogen levels were halved and 50% of the carcases had an ultimate pH at 24 h (pH_u) in the *M. longissimus dorsi* above 6.2. There was no clear indication that lactate, glucose, ATP or creatine phosphate levels were affected by fasting.

MORTALITY AND MORBIDITY

Deaths following transport are a well recognised hazard, particularly in calves. Thus in Britain Leech et al. (1968) showed that mortality in calves within three weeks of transport was 160% compared with similar calves kept on the farm of origin and Roy (1970) stated that it sometimes reached 215%. Rodenhoff and Schönherr (1971) found deaths in cattle carried by rail were double those for road transport.

Barnes et al. (1975) studied the effect of transport on 85 Holstein bull calves, aged 1 - 7 days when transported and weighing 36 - 54 kg. All were said to have received colostrum

Age (days)	1	2	3	4
Deaths (%)	35	30	10	0

The majority of deaths did not occur until 21 - 30 days of age. In an investigation of deaths of cattle during transport to an abattoir Zimmerman (1979) found a mortality rate of 0.012% in the period 1971 - 1976. Losses were higher in rail than in road transport and rose as distance increased up to 200 km, after which they declined. More deaths occurred in winter than in summer. The major causes of death were suffocation and asphyxia 38%, cardiac insufficiency 29%, hyperthermia 7%.

Jensen (1968) stated that the bovine respiratory disease (BRI) complex occurring in recently transported cattle was the single most important disease problem in the North American cattle industry. It is responsible for the deaths of approximately 1% of all cattle in United States feedlots. However, the economic loss from deaths is minor compared with the costs of prophylaxis, treatment and lost production.

In a study of morbidity and mortality in calves of various breeds, Staples and Haugse (1974) found that 41.6% of 3 249 calves over two weeks old when transported were ill during the following four weeks and 8.3% died in this period, whereas 60.3% of 1 769 calves under two weeks old when transported were ill in the following four weeks and 21.7% died. Calves of beef breeds showed the best survival figures. There was no significant relationship between morbidity, mortality and distance transported. Thus calves moved 50 miles had a four week morbidity of 56.77% and a mortality of 17.69%. The corresponding figures for journeys of 200 - 299 miles and for over 400 miles were 36.83% and 12.96%, and 45.32% and 11.98% respectively.

The suggestion that diesel exhaust gases contributed to respiratory disease in calves transported by road led Coleman and Sheldon (1976) to examine the performance of calves carried in trailers pulled by tractors with high and low exhaust pipes. They found that calves carried on the upper deck of trailers pulled by tractors with high exhaust pipes, i.e. approximately level with the trailer roof, showed an average daily liveweight gain during the first 70 days after transport 0.15 lb greater than calves carried on the lower deck. Conversely, calves carried on the upper deck of trailers hauled by tractors with low exhaust pipes, i.e. 12 - 24 inches below the trailer roof, had an average daily liveweight gain in the first 53 days after transport 0.14 lb less than the calves carried on the lower deck. The authors were unable to explain why calves on the lower deck of a trailer hauled by a high exhaust tractor subsequently had a lower rate of liveweight gain than those on the upper deck.

From an investigation of movements of feeder cattle by tractor-trailers in journeys of 1 000 miles, Camp et al. (1979) found a highly significant correlation between bodyweight and the development of shipping fever (Bovine Respiratory Disease: BRD). There is also an apparent effect on the incidence of this syndrome attributable to length of time in transit and/or weather conditions before, during and immediately after transit.

Irwin et al. (1979) compared the conditions under which cattle are transported in the United States and Australia. Distances traversed are roughly similar and Australian conditions are probably more severe, but the BRD complex is not recognised there. The greatest movement of feeder cattle in the USA occurs in September and October, the peak period for BRD, whereas in Queensland, for example, cattle are moved in May to July, a period of cool nights and warm, dry days. The disease syndrome has not been produced experimentally and the authors concluded that cattle in North America are exposed to greater climatic changes and, possibly, to more pathogenic viruses.

Early weaning, followed by feeding calves a concentrated ration for 30 days before transport (preconditioning), may reduce the incidence of the BRD complex, according to Cole et al. (1979), but it is not widely practised because of the high cost : benefit ratio and management factors.

INJURIES, INCLUDING BRUISING

McCausland et al. (1977) studied the occurrence of bruised stifle joints in the carcases of 16 400 young calves. Half the animals were affected and in 30% it was the only bruised region. The condition was usually bilateral and the gross and histological appearance was consistent with trauma, probably occurring in transit. Lameness amongst 411 15 month old Jersey heifers following transport was investigated by Dewes (1979). Within three weeks of arrival 36 were lame in one or both hind limbs. Excluding 5 clinically distinct cases, 31 did not go lame until 10 days or more after transport. Defects appeared in the claws, due to separation of the laminae from the wall over a localised area. This injury was considered to have occurred during the journey.

Susceptibility to bruising during transport was examined in five trials by Yeh et al. (1978). Steers, cows and a mixed group were included in each trial, which involved transport by lorry, followed by a longer rail journey. The weight of bruised tissue trimmed from the carcases of cows was significantly greater than that trimmed from steer carcases. Bruising of cows increased with the length of journey, but not in the case of steers. Wythes et al. (1979) recorded the effect on bruising and muscle pH of mixing cows and steers, or groups of strange steers, at various intervals before loading. The different intervals did not affect the degree of bruising occurring in carcases, but cows bruised significantly more than steers.

The relationship of fasting before transport to the degree of carcase bruising was studied by Dodt et al. (1979). Groups of 100 mainly Hereford and Shorthorn steers were fasted for 0, 24 and 48 h before transport, the second and third groups received no water in the last 24 h of their fasting period. The cattle were railed 245 km to the same abattoir.

Three separate trials were conducted, in the first two the
cattle were laired on arrival for 12 h, in the third for 36 h.
No food was given during lairing and the first two groups were
not watered, the third group had water available for 30 minutes
after unloading. The weight of bruised trim from cattle not
fasted before transport was significantly less than from the
fasted cattle, being respectively 0.48 kg unfasted, 0.99 kg fasted 24 h and 1.03 kg fasted 48 h. The weight of bruised trim did
not alter significantly with the length of fast before slaughter.

WEIGHT LOSS

Losses of body weight are probably the most significant
economic effect of transport. According to Thornton and Gracey
(1974) there appears to be a faster loss in the early stages of
transport than later, partly due to excretory loss, but tissue
loss starts early and continues uniformly for 96 h, after which
the rate decreases.

Self and Gay (1972) reported the results from 4 685 feeder
cattle, mostly weighing between 255 and 275 kg, transported by
road over distances ranging from 240 - 1 824 km (average 1 023 km)
in the United States. Cattle collected directly from ranches
showed an average loss of 7.2% bodyweight, while those collected
from auction markets averaged a 9.1% loss (P = < 0.05). Weight
loss increased by 0.38% for every 100 km of the journey. The
loss was recovered most rapidly in the spring and occupied
3 - 30 days, but 23 of 43 shipments regained the lost weight
within 7 days. No significant differences were observed
between calves and yearlings. Cattle taken directly from
ranches had the greatest shrink in summer (8.3%) and least in
the autumn (6.4%), while cattle collected from sale yards had
a uniform 8.9 - 9.2% loss. Loss of gut fill accounted for
46.7% of the total weight loss.

Chambers (1974) investigated weight loss and recovery in
three groups of heifers, averaging about 355 kg, in Rhodesia.
Two groups were transported 326 km by rail and one of these
groups was then reweighed and slaughtered 30 h after the start
of the journey. The other group was rested, with access to
water but no food, for a further 24 h before reweighing and

slaughter. The third control group was weighed and slaughtered without being transported. The cattle slaughtered on completing the journey had lost 12.88% of their initial mean liveweight. Assuming that the mass of ingesta was the same in all three groups at the start of the experiment, the ingesta lost in transit represented 11.8% of the liveweight and thus tissue loss accounted for 1.7%. The group rested for 24 h before slaughter had largely replaced this tissue loss.

Groth and Gränzer (1975) found that calves weighing 65 - 75 kg transported 223 - 265 km by lorry in 5 - 10 h lost 4.6 - 7.8% of their bodyweight. The same animals, when weighing 107 - 165 kg, were taken 266 - 310 km by lorry in 6 - 9 h and lost 5.5 - 6.6% of their bodyweight. In a comparison of combined rail and road transport with road transport only for the whole journey, involving 508 heifers, Schmalfuss and Käsebier (1978) concluded that high temperature and humidity in unsuitable rail wagons resulted in greater stress than road transport. Cattle hauled 280 km in 10.5 h by rail had an average weight loss of 24.9 kg/head, compared with 18.56 kg/head after a 350 km road journey in 16.5 h.

Camp et al. (1979) were unable to demonstrate a significant correlation between bodyweight and either weight loss or subsequent performance in feeder cattle hauled 1 000 miles by road. There was, however, a significant seasonal difference between weight loss and subsequent performance.

From a consideration of various marketing practices, including whether cattle are fed or watered before despatch for slaughter or after arrival at the slaughterhouse, the distance they are transported and the different purchasing arrangements made, Raikes et al. (1979) developed a procedure by which producers could compare bids from different sources for slaughter cattle.

IMPAIRED MEAT QUALITY

The commonest quality defect in beef attributable to transport is probably dark cutting, in which depletion of the glycogen reserves of the muscles results in failure of the pH to fall to normal levels *post mortem* . Prolonged stress as can result from mixing strange animals, e.g. in overnight lairing, from withholding food and water or exposure to marked diurnal temperature variations, will deplete muscle glycogen reserves. There appears to be general agreement with the comment by Ashmore et al. (1973) that dark cutting is the extreme example of the effects of pre-slaughter stress on meat quality and seems to result at least as much from emotional as from physical stress.

Nestorov et al. (1970) stated that muscle pH after transport had been shown to increase with the duration of the journey, but Carr et al. (1973) found that, while fasting cattle for 3 days decreased liver fat and glycogen, and reduced muscle water, it did not affect ultimate pH or water binding capacity. However, the pH_u of *M. longissimus dorsi* and *M. femoris profundus* was found by Shorthose et al. (1972) to be lower in steers rested and fed for 4 days before slaughter, after a 322 km journey, than in steers rested and fed for only 2 days.

In steers weighing 430 kg on average and transported 35 km Levantin et al. (1977) found it best from the aspect of meat quality to slaughter them on arrival. As the length of fasting rose, so did the pH_u. Thus steers slaughtered on arrival had an average pH_u of 5.85, those fasted 24 or 48 h had values of 6.15 - 6.2. In a subsequent trial by Levantin et al.(1978), similar cattle, but including bulls, were transported 45 km and slaughtered at 2, 24 or 48 h after arrival. The following pH values were obtained 2 h after slaughter.

TABLE 1

	2 h	24 h	48 h
Bulls	5.9 - 6.09	6.56 - 6.21	6.69
Bulls treated with chlorpromazine	5.97 - 6.03	5.58 - 5.71	6.54
Steers	5.94 - 5.99	6.27 - 6.41	• 6.17

In a survey of 682 commercial cattle slaughtered after overnight lairing in the early autumn (clear days and frosty nights) Duchesne (1978) found the condition in 8%. The incidence in young bulls treated similarly was 30%. Slaughter of bulls immediately on arrival at the abattoir can lessen the problem, but it is not an altogether satisfactory solution, as it depends on the reliability of transport and traffic conditions, factors outside the control of management. When young bulls were mixed in a previously empty pen at least 48 h before dispatch, dark cutting was reduced to about 5%, i.e. a level below that sometimes encountered with commercial cattle. This reduction applied even when the bulls were laired overnight so long as they were not penned with strange animals.

Observations on 423 steers penned in groups at a showground, transported by road for 5 h to an abattoir, where they were again penned in groups until slaughter, were reported by Grandin (1979).

TABLE 2

Number of animals	Treatment	Percentage dark cutting
80	Transported immediately and slaughtered on arrival at abattoir	0
45	1 night penned at showground, slaughtered on arrival	0
84	2 nights " " " " " " "	3.5
29	3 nights " " " " " " "	10
79	1 night " " " 1 night in lairage	13
109	1 night " " " resorted, 1 night in lairage	17

Reference has already been made to the finding by Fischer and Augustini (1980) that half the carcases from Simmental bulls transported for 4 h at different temperatures and then fasted 48 h before slaughter had a pH_u over 6.2.

The more frequent occurrence of blood splash in the carcases of pigs and sheep has led to a greater research effort into its causation in these animals than in cattle. Nevertheless, its sporadic occurrence in cattle can cause serious financial loss. Observations on its aetiology in this species were reported by Whittem et al. (1960) and Charles (1960) and knowledge of the causation of blood splash in cattle has advanced little in the last 20 years.

CONCLUSIONS

From the foregoing sections it is clear that, although many of the factors adversely affecting the physiology of cattle during transport are well recognised, their elimination or, at least, their reduction remains to be achieved. For example, few would disagree with the comment by Augustini (1976) that the greatest stress reaction is to loading and unloading. In discussing the humane handling of stock for slaughter, Kilgour (1978) remarked that a more complete ethogram (behaviour inventory) of the meat species would provide the key to better methods of handling each, thus reducing the stress and fulfilling humane requirements. He continued by commenting that application of existing knowledge, despite its limitations, would improve the general treatment of slaughter animals.

Behavioural studies of farm animals in transit, such as those reported by Grandin (1978) and by Jackson (1979) provide a basis for a better understanding of these problems and hence for their solution. Grandin (1980a, b) is a noteworthy pioneer in the practical application of ethological principles to the design of cattle handling facilities.

It is equally important that the training and supervision of staff responsible for the transport of cattle should receive constant attention.

REFERENCES

Ashmore, C.R., Carroll, F., Doerr, L., Tompkins, G., Stokes, H. and Parker, W., 1973. Experimental prevention of dark-cutting meat. J. anim. Sci., 36, 33-36.
Augustini, C., 1976. Electrocardiogram and body temperature measurements on pigs during fattening and transport. Fleischwirtschaft, 56, 1133-1137.

Barnes, M.A., Carter, R.E., Longnecker, J.V., Riesen, J.W. and Woody, C.O., 1975. Age at transport and calf survival. J. dairy Sci., 58, 1247.

Camp, T.H., Stevens, D.G. and Stermer, R.A., 1979. Transit factors affecting shrink, shipping fever and subsequent performance of feeder cattle. J. anim. Sci., 49, Suppl. 1,360.

Carr, T.R., Allen, D.M. and Phar, P.A., 1973. Effect of preslaughter fasting on some chemical properties of bovine muscle and liver. J. anim. Sci., 36, 923-926.

Chambers, P.G., 1974. Mass loss in slaughter cattle subjected to rail transportation. Rhod. vet. J., 5, 38-40.

Charles, D.D., 1960. Ecchymosis in the beef carcase. Australian vet. J., 36, 124-126.

Cole, N.A., McLaren, J.B. and Irwin, M.R., 1979. Influence of pre-transit feeding regimen and post-transit B Vitamin supplementation on stressed feeder steers. J. anim. Sci., 49, 310-317.

Coleman, J.D. and Sheldon, W.G., 1976. Beef Cattle Research in Texas, 1974-75, PR-3383C. Texas Agriculture Experiment Station, Bushland, Texas.

Crookshank, H.R., Elissalde, M.H., White, R.G., Clanton, D.C. and Smalley, H.E., 1979. Effect of transportation and handling of calves upon blood serum composition. J. anim. Sci., 48, 430-435.

Damron, W.S., Orr, C.L., Billingsley, R.D., Su, P.K. and McLaren, J.B., 1979. Rumen ciliate protozoa changes during weaning and marketing stress. J. anim. Sci., 49, Suppl. 1, 57-58.

Deutschmann, D., Kilb, F.E. and Grünn, E., 1976. Response of glutamic-oxalacetic transaminase, lactate dehydrogenase and aldolase in blood serum of cattle to transport stress. Arch. exptl. vet. Med., 30, 75-91.

Dewes, H.F., 1979. Transit-related lameness in a group of Jersey heifers. N.Z. vet. J., 27, 45.

Dodt, R.M., Anderson, B. and Horder, J.C., 1979. Bruising in cattle fasted prior to transport for slaughter. Australian vet. J., 55, 528-530.

Dotta, U., Abate, O., Gugliemino, R. and Giradi, C., 1976. Studies on the acid-base and electrolyte balance in cattle after transport. Atti Soc. Ital. Sci. vet., 30, 334-337.

Duchesne, H.E., 1978. The aetiology of dark-cutting beef. Ph.D. thesis, Bristol University.

Ewbank, R., 1973. Use and abuse of the term 'stress' in husbandry and welfare. Vet. Rec., 92, 709-710.

Fischer, K. and Augustini, C., 1980. Problem of dark-cutting beef. III. Effects of different transport temperatures and of food deprivation for several days. Fleischwirtschaft, 60, 469-473.

Grandin, T., 1978. Transportation from the animal's point of view. Technical Paper 78-6013, American Soc. agricultural Engineers, St. Joseph, Mich., USA.

Grandin, T., 1979. The effect of preslaughter handling and penning procedures on meat quality (steers). J. anim. Sci., 49, Suppl. 1, 147.

Grandin, T., 1980a. Livestock behaviour as related to handling facilities design. Inter. J. Study anim. Prob., 1, 33-52.

Grandin, T., 1980b. Observations of cattle behaviour applied to the design of cattle handling facilities. Appl. anim. Ethol., 6, 19-31.

Groth, W. and Gränzer, W., 1975. Influence of transport stress on the activity of GOT, GPT, LDH and CPK in the serum of calves. Zbl. vet. Med. A., 22, 57-75.

Groth, W. and Gränzer, W., 1977. Alterations of blood constituents due to transportation in fattening and early weaned calves. Deutsche Tierärztliche Wochenschrift, 84, 89-93.

Hails, M.R., 1978. Transport stress in animals: a review. Anim. regul. Stud., 1, 289-343.

Irwin, M.R., McConnell, S., Coleman, J.D. and Wilcox, G.E., 1979. Bovine respiratory disease complex: a comparison of potential predisposing and etiologic factors in Australia and the United States. J. Am. vet. med. Assn., 175, 1095-1099.

Jackson, W.T., 1979. Behavioural observations on farm animals in transit. Appl. anim. Ethol., 5, 291.

Jensen, R., 1968. Scope of the problem of bovine respiratory disease in beef cattle. J. Am. vet. med. Assn., 152, 720-723.

Kilgour, R., 1978. The humane handling of stock for slaughter with particular reference to procedures in New Zealand. Anim. regul. Studies, 1, 235-246.

Kriesten, K., Schmidtmann, W., Fischer, W. and Sommer, H., 1976. Influence of transport and sale stress on the concentration of total proteins, total lipids, glucose, creatine and electrolytes in the serum of stock bulls. Zbl. vet. Med. A., 23, 804-810.

Kriesten, K., Murawski, U., Egge, H., Fischer, W. and Sommer, H., 1978. Changes in serum lipids and fatty acids of stressed bulls. Zbl. vet. Med. A., 25, 207-221.

Leech, F.B., Macrae, W.D. and Menzies, D.W., 1968. Calf wastage and husbandry in Britain, 1962-63 H.M.S.O., London.

Levantine, D.L., Fomichev, J.P. and Afanasjev, E.S., 1977. Influence of transportation stress, fasting and aminazine on the live and carcass mass losses and chemico-physical characteristics of the meat of calves and steers. 23rd Europ. Meet. Meat Res. Wkrs. Vol. I, Paper B3. All-Union Meat Research Institute, Moscow.

Levantine, D.L., Fomichev, Y.P., Afanasjeva, E.S., 1978. Effect of length of resting period before slaughter of young bulls and steers and treatment with aminazine (chlorpromazine) tranquillizer during transport on physical and chemical processes in meat during storage. 24th Europ. Meet. Meat Res. Wkrs. Vol. I, Paper 10. Bunderstalt für Fleischforsching, Kulmback.

McCausland, I.P., Austin, D.F. and Dougherty, R., 1977. Stifle bruising in bobby calves. N.Z. vet. J., 25, 71-72.

Mouthon, G., Longin, C. and Magat, A., 1976. Effect of transportation on the blood pH, pCO_2 and pO_2 in young beef cattle. Bull. Soc. sci. Vét et Méd comparée de Lyon, 78, 333-336.

Nestorov, N., Tomov, T. and Krestev, A., 1970. A study on transport stress in cattle and conditions for its manifestation. 16th Meet. Europ. Meat Res. Wkrs. Vol. I, Paper A24. Meat Technology Research and Project Institute, Sofia.

Pérez-Fernández, L.F. and Willoughby, R.A., 1976. Study of the changes in adrenocortical activity as a measure of stress in cattle. Veterinaria Mex., 7, 3-8.

Raikes, R., Sieck, G.M., Self, H.L. and Hoffman, M.P., 1979. Weight loss of fed steers and marketing decision implications, Special Report 82. Iowa State University, Ames, Iowa.

Rodenhoff, G. and Schönherr, S., 1971. Causes of damage in animal transport. Tierärztliche Umschau, 26, 4-6 & 8-9.

Roy, J.H.B., 1970. The Calf. Vol. 2, Nutrition and Health, pp. 111-114. Iliffe, London.

Schmalfuss, R. and Käsebier, L., 1978. First experimental results regarding the transport of heifers by various means. Tierzucht., 32, 182-184.

Self, H.L. and Gay, N., 1972. Shrink during shipment of feeder cattle. J. anim. Sci., 35, 489-494.

72

Shorthose, W.R., Bouton, P.E. and Harris, P.V., 1972. The effects on some properties of beef of resting cattle after a long journey to slaughter. Proc. Australian Soc. anim. Prod., 9, 387-391.

Simensen, E., Laksesvela, B., Blom, A.K. and Sjaastad, O.V., 1980. Effects of transportations, a high lactose diet and ACTH injections on the white blood cell count, serum cortisol and immunoglobulin G in young calves. Acta vet. scand., 21, 278-290.

Staples, G.E. and Haugse, C.N., 1974. Losses in young calves after transportation. Br. vet. J., 130, 374-379.

Thornton, H. and Gracey, J.F., 1974. Textbook of Meat Hygiene, 6th Ed., pp. 422-434. Baillière Tindall, London.

Velinov, P. and Enev, E., 1978. Ultrastructural changes in the meat of calves slaughtered without pre-slaughter rest. 24th Europ. Meet. Meat Res. Wkrs. Vol. I, Paper All. Bundestalt für Fleischforschung, Kulmbach.

Whittem, J.H., Letts, G.A., Rideout, F.C. and Frith, M.R., 1960. Ecchymosis in beef cattle. Australian vet. J., 36, 122-124.

Wythes, J.R., Gammon, R.H. and Horder, J.C., 1979. Bruising and muscle pH with mixing groups of cattle pre-transport. Vet. Rec., 104, 71-73.

Yeh, E., Anderson, B., Jones, P.M. and Shaw, F.D., 1978. Bruising in cattle transported over long distances. Vet. Rec., 103, 117-119.

Zimmermann, I., 1979. Data on the transportation of animals with respect to some meteorological factors. Magyar Allatorvosok Lapja, 34, 191-195.

DISCUSSION

E. Kallweit *(Federal Republic of Germany)* You mentioned CPK being increased by transportation with a rapid decrease afterwards. Can you tell us how long it takes to decrease to the normal level? The reason for my question is that this is a rather slow process in pigs, with CPK increasing over 12 - 24 h to a maximum and then levelling out again, and so I was surprised to hear that it occurs much faster in cattle.

T.M. Leach *(UK)* I could let you know later.

P.V. Tarrant *(Ireland)* I would just like to make a comment on the question. We found that the level of CPK in serum from a mixed group of bulls increased between ten- and one hundredfold within six hours of the time of mixing and the level did not fall back to resting value until between 48 and 72 h after the start of mixing. Mixing terminated after 6 h and the animals went back to their original stalls.

D.B. Stephens *(UK)* You mentioned the training of staff to handle livestock. Can people be trained to handle cattle or must they be born with an innate ability? My opinion is that it is an innate factor which cannot be learned.

T.M. Leach That can be said of any occupation. There are two schools of thought, one that people are born to an occupation and one that you can be trained to it. I would have said that if a man had a reasonable intelligence and a basic sympathy with animals, training would be possible. Our Agricultural Training Board in Great Britain does a great deal of training and I think they are successful. I would not say that they are always successful but standards afterwards are nearly always better than they were before.

W. Sybesma *(The Netherlands)* What do you consider is the factor which threatens welfare the most? You said that the immunogenic system is threatened. Would you propose to treat the animals after transport?

T.M. Leach The particular losses which I mentioned involved very young animals, so the answer may be to avoid transport in the first few days of life.
This would seem to be important.

W. Sybesma Is mortality due to the new environment or to the transport?

T.M. Leach Presumably it is because of the immunological incapacity of the
very young animal. In that paper the young animals were all said to have
had colostrum, but it is doubtful that day old calves had received much, and
in any case, we do not know how they were responding. They would only have
a short umbrella cover from colostrum. I think that their immunity was chal-
lenged beyond its capacity. It is difficult to see how one would overcome
this apart from suggesting that they were not transported until they were a
little older.

W. Sybesma You would not like to treat them with corticosteroids or something
like that, to enhance the response?

T.M. Leach I do not have the experience, but I would have thought that it
would be a rather expensive way of tackling the problem, if you could avoid
this simply by holding the animals a little longer before transporting them.
It is flirting with danger to move them in the first two or three days of
life.

W. Sybesma It depends on the situation. Sometimes it is necessary to move
them.

T.M. Leach In that case I do not know whether corticosteroids would work or
not. I have not seen any published work on this.

E. Wagner (Luxembourg) You mentioned that the enzymes and cortisol and so
on were increased by transport. My questions are (1) do they reach pathological levels and (2) can they be considered as criteria for the measurement
of animal welfare?

T.M. Leach They can reach pathological levels, but how you would relate that
to animal welfare I am not sure, except to say that if the animal was in this
state it was unfit or had been made unfit by the stress of transport. As
Dr. Lister said, it is often very difficult to use this sort of information
in a direct welfare argument.

J.E. Melville (Australia) I should like to make a comment on the question
of looking at blood parameters as indicators of stress. A fair amount of
work has been done on the change in blood parameters due to the stress of
taking blood samples from live animals, particularly on pack cell volume and
some of the red and white cell counts, and I wonder sometimes just how good
some of these parameters are.
     Secondly I should like to turn to the question of weight loss and water
deprivation, and the regaining of weight if the animals are given water. Has
anyone looked particularly at the effect of giving water plus electrolytes?

T.M. Leach Not that I know of, although it seems a reasonable thing to do.
I have certainly not seen any work on that in cattle, but this is the sort
of thing I think one might well consider.
     Your other point is mentioned in one or two of the papers I quoted,
and our own work at the Institute has shown this when we sample sheep and
pigs regularly. The values fall when the procedure has become familiar, and
this does limit, to some extent, the value of single time tests when the
animal is in transit. It is a factor which must not be forgotten.

E. Kallweit I would like to comment on the first part of the question. I
think that it always depends on the kind of characteristics which you are
going to measure in the blood. If they are easily changed by any kind of

stress, then the taking of blood samples in the ordinary way should be avoided. To work in a scientific way, you should work with permanent cannulae and catheters and get the pig used to the procedure of blood sampling in the way which I shall speak about in my paper.

P.J. O'Connor *(Ireland)* Having spent many years in meat inspection I am well aware of the damage which can be caused by cattle with horns. We made a national law which states that no cattle can be exposed for sale if they have horns. I would like to know if this sort of practical measure has been taken elsewhere. Dr. Leach may be able to say whether this is a law in the UK at the moment.

T.M. Leach We have no such law. The Australians have done work and have shown the damage which horns can cause in comparative groups and when mixing horned and polled cattle together. In Britain we do not have a law, but horns have almost disappeared, so it is not a problem.

D.B. Stephens I thought that in the UK we were not allowed to transport horned cattle in the same truck as unhorned cattle.

T.M. Leach They must separated.

A. Cuthbertson *(UK)* We have been talking about pigs and cattle so far, and we have rather a lot of sheep in the United Kingdom. I am wondering to what extent the evidence which has been collected on cattle is appropriate to sheep, as they have a ruminal function like cattle? For example, can they withstand longish periods without feed and water?

T.M. Leach There are other problems with sheep. I did not touch on blood flow. Blood splash in cattle is not a major problem whereas it is sometimes in sheep, and on occasions this is undoubtedly related to transport. I think that most of the biochemical reactions of cattle are probably fairly similar to those of sheep during transport.

PHYSIOLOGICAL RESPONSE OF PIGS TO TREADMILL EXERCISE USED AS A STANDARDISED STRESS

E. Kallweit

Institut für Tierzucht und Tierverhalten,
FAL, Mariensee, 3057 Neustadt 1,
Federal Republic of Germany.

ABSTRACT

To investigate physiological reactions in slaughter pigs following stress, as' for example in transportation to the slaughter plant, groups of animals were exercised on a treadmill for 5 min each at a speed of 1 m/sec. One animal in each of 6 groups was cannulated and at fixed time intervals thereafter for a total period of 5 h blood samples were taken from the cannulated pig. At each of these sampling points, another pig of the group was slaughtered. Samples of exsanguinated blood were taken as well as muscle samples and analysed.

The results indicate significant effects of treadmill treatment on the acid-base status, LDH activity, cortisol concentration and lactate in blood. During the recovery period most of the parameters returned to the original level within 60 - 120 min with the exception of LDH.

Changes mentioned above due to stress were similar in catheterised and in slaughtered pigs. In the latter, however, the reactions were more pronounced.

Meat quality and substrates in the muscle were not affected significantly by the treatment, but there were tendencies apparent shortly after stress towards quality deficiencies. In the liver, however, glucose and glycogen were decreased and did not recover within the experimental period of 5 h. Despite the fact that many of the experimental animals were exhausted after treadmill treatment, the glycogen content in the muscle remained unchanged and the exhaustion must have been due to excitement rather than to physical stress.

It can be concluded that pigs should not be slaughtered directly after arrival at the slaughter plant, but should be rested for at least 1 - 2 h, if they are stressed only by transport. In case of real physical exhaustion associated with glycogen depletion, even more time should be allowed for recovery.

INTRODUCTION

Federal German legislation provides that pigs must be rested before slaughter. There have been various proposals for the optimal period for such recovery before slaughter and the survey by Augustini and Fischer (1981) lists periods ranging from none to 48 h for resting animals. The problem is how to evaluate adequately the process of recovery from stress induced, for example, by transport or handling animals before slaughter.

Stress in the pig results in a number of physiological changes including elevated body temperature, a change in the acid-base balance, and increased serum enzyme levels (Haase, 1972; Mäder, 1974; Schmidt, 1980). In the present study we have measured several physiological parameters for up to 5 h following treadmill exercise which we have used as a readily reproducible form of stress. By use of indwelling jugular cannulas, we have avoided the added stressor of blood sampling and this has allowed a comparison of the return to normal of physiological parameters with the changes in meat quality in pigs slaughtered at comparable times after the exercise stress. Nevertheless excitement caused by any departure from normal activities may act as a stress inducer in individual animals and while considerable attention has been paid to obviate this factor it remains as a source of variation difficult to control. A comparison then between catheter blood and exsanguinated blood samples provides an assessment of differences between exercise and exercise coupled to the stress associated with slaughter of a group of animals.

MATERIALS AND METHODS

The animals used were German Landrace pigs of 105 kg liveweight and, to avoid the more obvious genetic influences, the animals in each of the 6 groups of 11 pigs used were bred from sows who were littermates mated to the same sire. One pig from each of the 6 groups was cannulated via the *Vena cava cranialis* and was habituated to the procedure of blood sampling for two weeks. Body temperature was measured and blood samples were collected at rest before exercise (control), directly after exercise (= 0 min) and subsequently at 9 time intervals as shown in the results. The exercise regime for 5 min was on a treadmill operated at a speed of 1 m/sec. The state of fatigue of the animals was visually appraised.

The remaining animals in each group were exercised on the treadmill identically after which one animal was slaughtered at each of the time intervals corresponding to those at which blood was withdrawn from the cannulated control animal. At the time of slaughter samples of the exsanguinated blood were collected. The cannulated pig was rested for 18 - 24 h at which time it was slaughtered as the control slaughter pig without exercise.

Meat quality characteristics of colour brightness and pH were measured at 10 min, 30 min and 24 h *post mortem* while liver samples were analysed 20 min *post mortem*.

Measurements of P_{O_2}, P_{CO_2}, cortisol, lactate, glycogen and glucose were taken by standard methods (Fehrentz, 1976). Results given are the mean (± SD) of six animals and the significance was determined by analysis of variance.

RESULTS AND DISCUSSION

The results of measurement of a variety of physiological parameters which respond to stress in the pig are shown in Figures 1 - 11.

To summarise the results a number of points can be made.

1) In all instances the physiological response to exercise was more extravagant and in general persisted longer in the slaughtered animals than in the quiescent control (cannulated pigs). This emphasises that there are two separate stress mechanisms operating, one associated with exercise and the other a slaughter stress related to the stimulus of the proximity of the slaughter activity. An example of this is the cortisol levels and the haematocrit (Figures 1 and 2).

2) There are wide variations in response as indicated by the large SD in the slaughtered animals, which points to a wide variation between animals in excitability, associated with slaughter. This tendency is less marked in direct responses to exercise seen in the catheter samples. Once again this is best illustrated by reference to the cortisol levels.

3) Normalisation of physiological response to exercise is apparent in the main between 1 and 2 h. Thus the pH and P_{O_2} (Figures 3 and 4) returned to control values within 30 min while body temperature and lactate remained elevated for 120 min (Figures 5 and 6). A major exception was lactate dehydrogenase (LDH) (Figure 7) which had not peaked 5 h after exercise. It is therefore suggested that measurement of serum LDH levels may provide insight into prior exposure to stress.

78

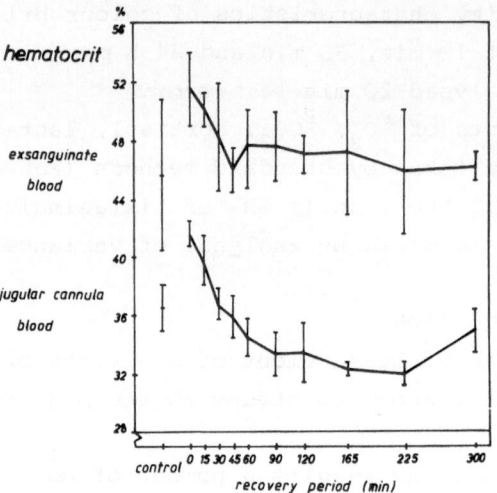

Fig. 1.   Haematocrit of cannula and exsanguinate blood following treadmill
exercise.

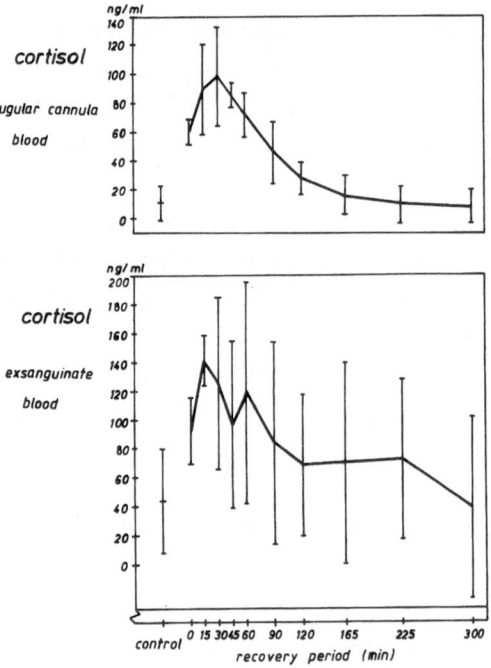

Fig. 2.   Cortisol of cannula and exsanguinate blood following treadmill
exercise.

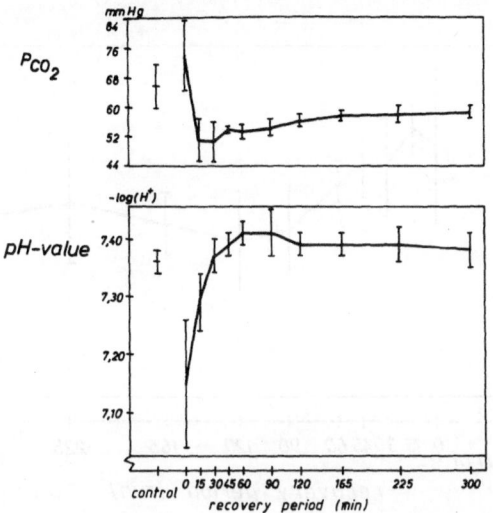

Fig. 3. P_{CO_2} and pH of cannula blood following treadmill exercise.

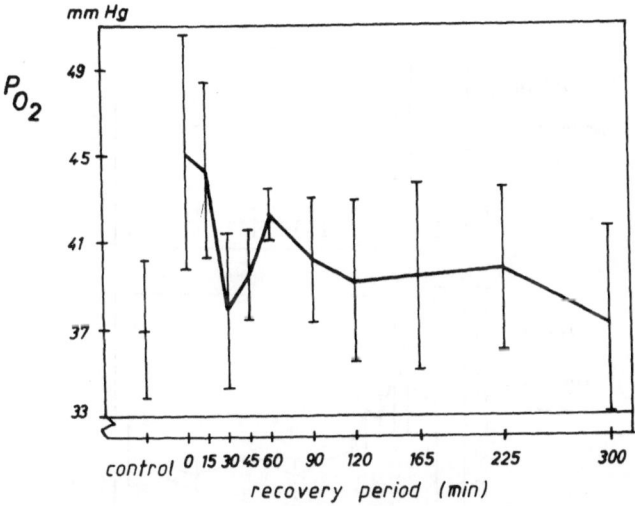

Fig. 4. P_{O_2} of cannula blood following treadmill exercise.

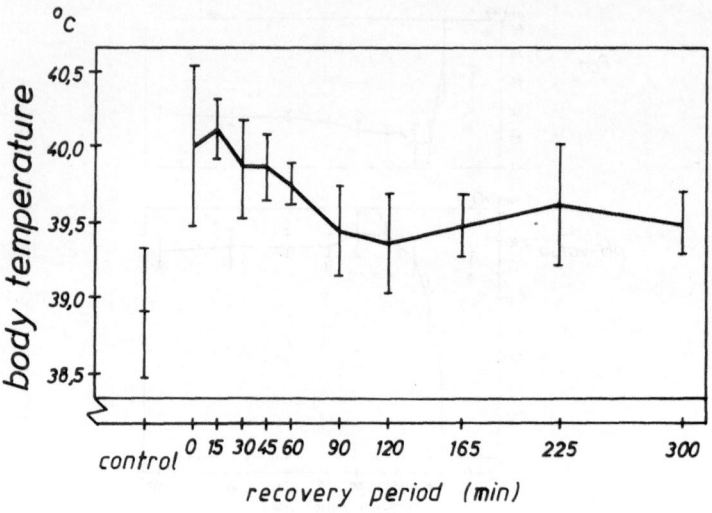

Fig. 5. Body temperature of catheterised pigs following treadmill treatment.

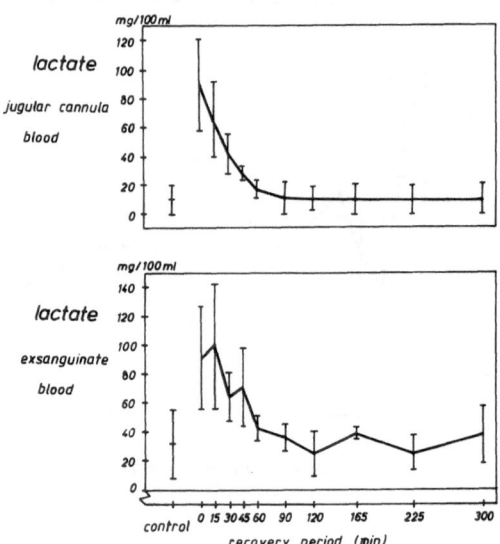

Fig. 6. Lactate of cannula and exsanguinate blood following treadmill exercise.

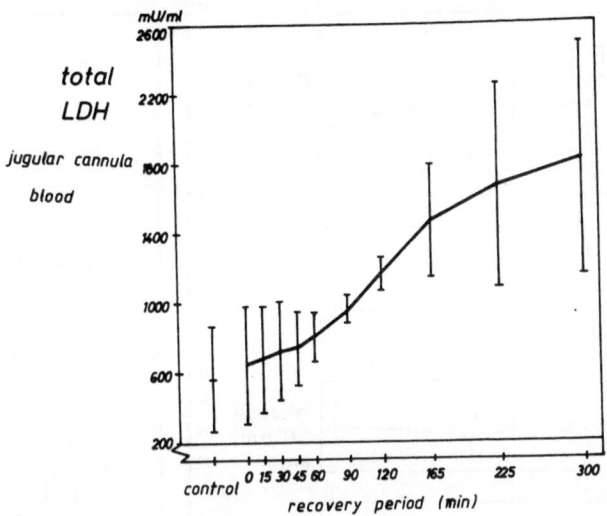

Fig. 7. LDH of jugular cannula blood following treadmill exercise.

Results of measurement of *post mortem* pH, glycogen, glucose are shown in Figures 8 - 11 from which the following considerations emerge.

1)   There was no extensive energy depletion to the point where meat quality was seriously impaired by the effects of stress (Figures 8 and 9), although immediate slaughter following exercise stress gave indications that it may well be less beneficial than following a period of recovery. This can be inferred from Figures 10 and 11 for colour brightness and pH values in the *M. longissimus dorsi*.

2)   Some animals appeared to be fatigued during treadmill exercise and increases in body temperature (Figure 5), cortisol (Figure 2) and haematocrit (Figure 1) undoubtedly reflect this state. However these values had normalised in 1 to 2 h.

3)   With both exercise stress and stress related to the excited state of the animals even in the extreme responses, a holding time of 2 to 3 h was sufficient to restore a near normal *post mortem* product.

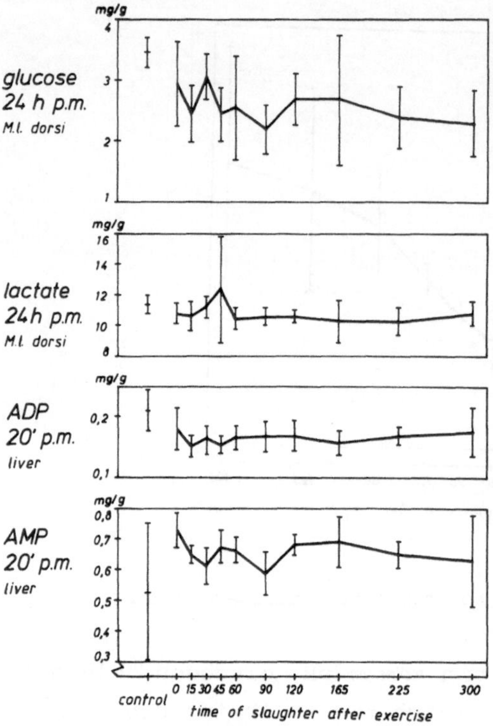

82

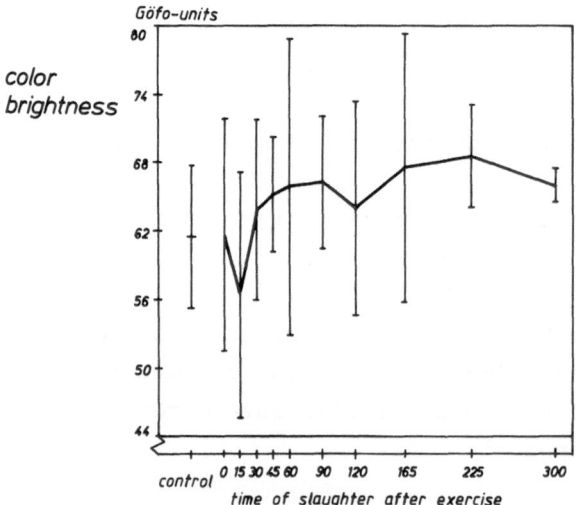

glucose 24 h p.m. M.l. dorsi

lactate 24 h p.m. M.l. dorsi

ADP 20' p.m. liver

AMP 20' p.m. liver

glycogen 20' p.m. liver

glucose 20' p.m. liver

Fig. 8. Glucose and lactate of *M. longissimus dorsi* samples 24 h *post mortem* and ADP and AMP in liver samples 20 min *post mortem* of slaughtered pigs following treadmill exercise.

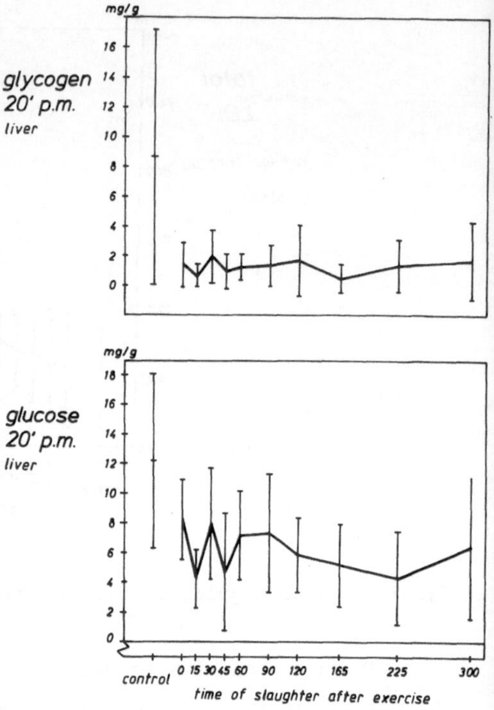

Fig. 9. Glycogen and glucose of liver samples 20 min *post mortem* of slaughtered pigs following treadmill exercise.

color brightness

Fig. 10. Colour brightness (Göfo-values) 24 h *post mortem* of *M. longissimus dorsi* of slaughtered pigs following treadmill exercise.

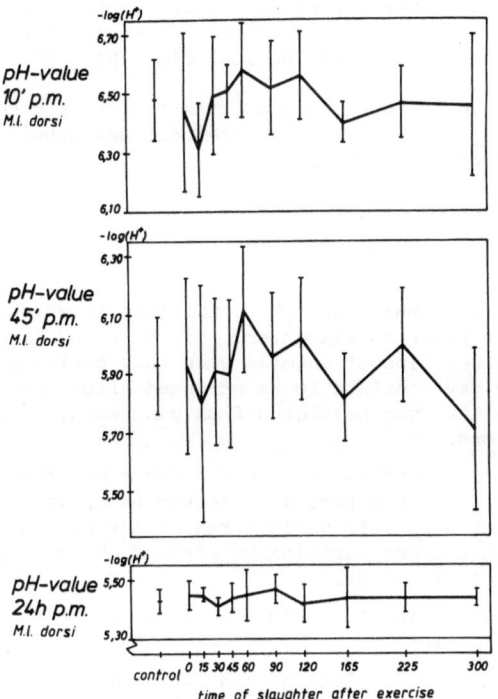

Fig. 11. pH in *M. longissimus dorsi* at 10 min, 45 min and 24 h *post mortem* of slaughtered pigs following treadmill exercise.

In conclusion it is axiomatic that more extreme cases involving prolonged transportation and stress in which glycogen stores are severely depleted will require a correspondingly longer time to recover and at all events slaughter directly following transport should be avoided.

REFERENCES

Augustini, Chr. and Fischer, K., 1981. Behandlung der Schlachtschweine und Fleischbeschaffenheit - eine Felduntersuchung. Fleischwirtschaft 61, 775-785.
Fehrentz, G., 1976. Belastungsreaktion und Fleischbeschaffenheit beim Schwein in Abhängigkeit von der Zeit nach einer Belastung. Dissertation, Gießen (Ph. D. Thesis).

Haase, S., 1972. Untersuchungen zur Erfassung des Adaptationsvermögens
    beim Hausschwein. Schriftenreihe des Max-Planck-Instituts für
    Tierzucht und Tierernährung Mariensee/Trenthorst, Heft 57/1972.
Mäder, H.-P., 1974. Beziehungen zwischen Belastungsreaktionen und der
    Fleischbeschaffenheit beim Hausschwein. Dissertation, Berlin (Ph.
    D. Thesis).
Schmidt, U., 1980. Die Eignung des Halothantests zur Erkennungder
    Belastbarkeit und Fleischbeschaffenheit des Schweines. Landbaufor-
    schung Völkenrode, Sonderheft 50.

DISCUSSION

A. Cutherbertson *(UK)* What were the conditions in which the animals were
handled immediately prior to slaughter? If you think of commercial practice,
I suspect that the handling of animals when they arrive at the slaughterhouse
may be poor. They may continue to be stressed after the journey and not
recover in the way that you predicted from your work, done, perhaps, under
laboratory conditions.

E. Kallweit *(Federal Republic of Germany)* This was our goal. We wanted to
standardise handling, so the pigs were rested as optimally as possible after
the treadmill. They were left undisturbed as far as possible until the time
of slaughter, so that there was minimum stress. They only had to walk about
20 m to the place of slaughter.

A. Cuthbertson How far is it possible to extrapolate these findings to variable commercial practice?

E. Kallweit We wanted to exclude as many variables as we could, of those we
could control, and this is why we applied this type of stress.

G. von Mickwitz *(Federal Republic of Germany)* What do you think is the best
resting time for good slaughter condition?

E. Kallweit If the pig is not very severely physically stressed, if it is
only excited, then about two to three hours should be sufficient. Even in
the treadmill treatment the reaction of the pigs varies. When they get onto
the treadmill for the first time the reaction is much more pronounced than
in later treatments. A training effect is seen. If pigs could be trained even
once to transportation or loading, it would avoid a lot of deaths. This is
already done by some breeders, e.g. if a breeder wants to take a boar to an
auction, he will train the boar to enter the truck. Where the animals are
really expensive, training is done.

W. Sybesma *(The Netherlands)* I should like your comments on the fact that
if there is a period of waiting before slaughter, there is also an accumulation of salmonella. I think you should consider that if you want to
recommend this procedure.

E. Kallweit I would not like to comment on that but would ask one of the
veterinarians. Our veterinarians suggest that, because they say that if an
animal comes to slaughter with an elevated body temperature it is not ready
to be slaughtered. It may be a problem, but I do not have a solution for it.

W. Sybesma You can lower the temperature by showering.

E. Kallweit Sure, but you need some recovery time. It may be an hour or
two, at least.

G. von Mickwitz We need a resting time, and therefore I think it is very important to know what that time should be. I think that 12 h may be a problem because of salmonella, but if that is so, you should look to see where that salmonella comes from. Cutting down the resting time is not the answer to the problem of salmonella.

I do not think it is possible to lower the temperature by using showers. We have found this in the course of experiments on many thousands of pigs. If the pigs have a temperature of 42°C, for example, the shower will only lower that by 1°C. The advantage of the shower is the effect it has on the nervous system.

D. Lister *(UK)* I am sure that for most practical purposes it is more important to reduce the amount of time an animal spends during transport, rather than to think about how many hours it needs to recover. I think there are too many problems. Dr. Sybesma raised one concerning salmonella. I think that the risk of the animal encountering even more stress during lairage is as great and I would be very reluctant to see a proposal to extend lairage times to accommodate recovery.

N.J. Nielsen *(Denmark)* Like Dr. Lister, I would ask whether you can tell anything about the resting time before slaughter without mentioning the distance over which the animals have been transported. In Denmark we have found that there is a close relationship between the time of last feeding, the time of transportation and the time of resting in pens. In Denmark the distances between farms and factories are, in general, short and we have found, in looking at meat quality and PSE problems, that it is necessary to have a resting time of about 4 h after a short transport distance.

E. Kallweit I think that the statement I made cannot be generalised for all conditions. We tried to keep the conditions as constant as possible by just having one kind of stress, to see what happens under those conditions and how fast the pig will recover. If you put additional variables into the system, it cannot give the answer for all conditions.

To come back to Dr. Lister's remark, if the resting conditions are bad, I think it would be better to have no resting time. If the conditions are good, and the water put on the pigs is not a shower, but comes from a sprinkler, then that will make a difference. Simply pouring water over the pigs is more likely to stress than to quiet them, and all these factors have to be taken into consideration. If the pigs are rested and then you get them up in a wild way, all the effort will be wasted.

P.J. O'Connor *(Ireland)* I would like to comment on the applications of minimum periods of resting. Our law puts the responsibility on the veterinarians who carry out the *ante mortem* inspection to decide whether the animals require further rest. Some years ago the decision was made that there should be a minimum time of 4 h. That had to be abandoned because we found that it was not practical. It placed an onus on people to control the intake of pigs, to segregate them and keep records of arrival, and if you are going to apply a law like that you are tying yourself up in red tape. We decided to abandon that minimum period. Allowing the veterinary surgeon to look at the pigs and decide is better, because, as our colleague from Denmark said, some are coming a very short distance. The 4 h rule meant that those pigs would have to rest for 4 h when they did not need it. It is better to put the onus of decision on the veterinary inspector.

E. Kallweit I think that the distance of transport is no excuse. Even within our institute we have transported pigs over 10 km and four out of twenty were dead on arrival, under unfavourable conditions. Loading and unloading are the marker points, so the distance itself does not play such an important part.

T.M. Leach *(UK)* We have certainly found, comparing pigs that have been hauled 15 miles with pigs that have been hauled 150 miles that the short haul pigs were more stressed. This would support the Danish observation, because it is true to say that they recover on the lorry if their loading rate is right and the driving is good. We have found in the lairage situation with good quality pigs hauled for 4 - 6 h that we could show no advantage in lairing for more than 1 - 2 h in terms of pH fall or muscle probe quality. We felt that 1 - 2 h was therefore probably adequate rest.

J.G. van Logtestijn *(The Netherlands)* A lot of work was done in our country some ten years ago by Verdijk on the influence of the resting period. He found that in normal circumstances a resting period of 2 - 4 h was optimal and the percentage of PSE carcases decreased sharply. With resting periods of 4 h or more the percentage of PSE carcases increased very sharply. He concluded that a period of 2 - 4 h was, in general, the most optimal resting period.

E. Kallweit There is agreement on this point, I think, and one reason may be that after pigs have rested for about 2 h they have recovered and get up and start fighting again, so that the process is reversed.

P.V. Tarrant *(Ireland)* I wonder if Dr. Kallweit would tell us which physiological indicators he would favour and whether any of these can be looked at in slaughter blood as opposed to live samples?

E. Kallweit I shall start with the second part of your question, because I have never thought extensively about the first part. I think you could measure the plasma enzyme activity in slaughter blood because this may give indication of a long term stress, but the rest of these parameters change so rapidly that I doubt whether they can be used. Most of them can only be analysed in a laboratory so that there would be practical problems. The exception may be body temperature, because if body temperature is high the pig is suffering.

P.V. Tarrant Would you accept that a cortisol measurement on slaughter blood would relate reasonably well to a cortisol measurement taken immediately pre-slaughter?

E. Kallweit This cannot, I think, be analysed on the spot. You need an answer right away if you are going to ask a question like that. There is a correlation even though the variation is greater. If the final analysis is sound then it may work.

G. von Mickwitz If you have normal transport, pigs should have recovered within 1 h. Clinical signs such as heart rate should have gone down and temperature and respiration rate should have reached normal values. The arterioles on the ears should be normal and not full. This is after 1 h. All pigs transported in bad conditions will show these signs for more than 2 - 3 h.
  To return to the question of the resting period, if you have a higher percentage of myopathy, that means more PSE, it is best to stun and kill them as soon as possible. All slaughterhouses do this. They want to kill them and put them into the cool box so as to keep the water until the meat is bought. Then you will see the water.

E. Kallweit I should just like to make one comment on myopathy, or else the discussion will take a wrong turning. In my opinion, myopathy tissue looks very different from PSE tissue. Over the last few weeks I have seen many examples of myopathy in pigs. We measured the pH and found that it was up to 6.2 or even higher, whereas in the rest of the tissue pH was normal, i.e. about 5.4 - 5.6. The actual status of myopathy is a little different from PSE.

BEHAVIOUR OF CATTLE DURING TRANSPORT AND PENNING
BEFORE SLAUGHTER

F.J. Kenny[1] and P.V. Tarrant

[1]Department of Psychology, University College,
Dublin 4, Ireland.

Meat Research Department, An Foras Taluntais, Dunsinea,
Castleknock, Co Dublin, Ireland.

ABSTRACT

The behaviour of cattle during transportation and the associated handling and penning procedures is reviewed. The activities of cattle during transportation include heightened activity immediately after loading and characteristic standing and lying behaviour in moving and stationary vehicles. The available transport data come almost exclusively from rail as distinct from road travel. During the handling of cattle, certain characteristic behavioural patterns are also displayed. These are identified and their influence on personnel and relevance to animal welfare discussed. The available information on behaviour in lairage before slaughter concerns the problems of mixed penning and horned cattle, and also feeding and resting behaviour.

It is concluded that behavioural data relevant to the preslaughter period are scarce and much more basic and applied research is necessary to provide a proper assessment of the welfare of cattle at this time.

INTRODUCTION

Preslaughter management of cattle is of concern from the welfare viewpoint because it is a common and final experience in the lives of virtually all cattle and because there is evidence, mainly from statistics of mortality and carcase and meat quality defects, that problems of well-being of animals exist. Change of ownership frequently occurs and this, together with the imminence of slaughter tends to reduce the level of care. The most powerful stimulus towards improvement in care of slaughter stock comes from the need to avoid financial losses. These losses occur in considerable amounts and in any case represent only the most obviously inadequate situations. The study of animal behaviour provides a direct and positive means of assessment of well-being.

In any study of animal welfare it is important to outline the behaviour of the animal under study in its normal 'complete' environment. Then, the behaviour of the animal in the system under study can be compared to this. Absences of specific behaviours, or decreased or increased frequencies or distortions can be isolated and an informed opinion given as to their importance.

In the case of the transport of livestock we are dealing with a situation which has no comparable equivalent occurring naturally. The animal is coming from a relatively tranquil and undisturbed environment. On his journey he experiences overcrowding, unstable floor surfaces and mixing with strangers, often for the first time. His innate characteristics, past history and the features of the novel situations interact to determine how he copes. The present paper reviews much of the available literature on this subject which will be considered under the following areas: behaviour during transport; behaviour during handling, loading and unloading; and behaviour in lairage.

THE BEHAVIOUR OF CATTLE DURING TRANSPORT

Behaviour during long distance (4 - 5 days) rail transport was studied by Bisschop (1961) and Sutton et al. (1967). Anderson (1973) also studied behaviour during road and rail transport in relation to carcase bruising. No reports concerning routine, short-haul road transport were found in the literature reviewed. Behaviour during transport may be conveniently categorised as follows: behaviour after loading; standing in transit; lying down in transit; and behaviour in stationary wagons.

Behaviour after loading

Bisschop found two different sets of behaviour occurring in the interval between loading and departure. In one case, where 1½ h elapsed between loading and departure, the steers (of mixed breed) were very nervous and restless, showed trembling especially of the hind-flanks and kept on butting each other and changing their positions in the trucks. They defaecated and urinated so copiously that they soon stood inches deep in faecal slush. This state lasted for the first 4 h after departure although by then the steers had stopped trembling. During the following 8½ h the animals settled down. They stood resting, mostly with their heads low and some with their eyes

closed and apparently asleep. Urination and defaecation stopped.
It appeared as if intestinal contents and free body moisture
had been drained.

In the other two cases studied, (one consisted of cattle
of mixed breeds and the other of Afrikaner-type cattle), 3 h
elapsed between loading and departure. No animals showed
nervousness, trembling or more than normal defaecation and
urination. They did however mill about.

Bisschop offers no explanation for this difference. However, the first group travelled 8 h to the railhead and seemed
to have spent just one night there. The other groups travelled
2½ days and 5½ days to the railhead and spent 19 and 29 h there,
respectively. It is possible that the first group had not
sufficient time to complete formation of a social hierarchy,
in view of the large group size of 110 animals, and that this
was being completed on the wagons. Sutton et al. (1967) noted
the behaviour after loading of Afrikaner-type oxen from a single
herd. Initially they were restless and anxious. A high state
of alertness was suggested - they held their heads high, well
above the level of the back and the eyes were prominent. This
stage of unsettled behaviour lasted only 30 min, probably due
to the established nature of the group.

Standing behaviour

Groups of cattle tend to settle down soon after departure
(Bisschop, 1961; Sutton et al., 1967), although occasional
groups exhibit continued nervousness as described above
(Bisschop, 1961). While they are in motion, animals stand
parallel to each other and at right angles to the direction
of travel of the train, facing indiscriminately in either
direction (Sutton et al., 1967; Anderson, 1973; Bisschop, 1961).
This position appeared to be the best to counteract the sideways sway of the truck when the train was moving (Sutton et
al., 1967). The head was held with the poll at the same height
as the back or lowered with the muzzle about 15 cm above the
floor of the truck. The animals only raised their heads if
they were disturbed, some gave low moans and others ground
their teeth (Sutton et al., 1967).

Standing during transport introduces a fresh dimension
of postural adjustment in order to stay upright. Body weight
at rest could be expected to be evenly spread over four feet
(Dewes, 1979) and Sutton and co-workers suggested a similar
weight distribution during motion. Dewes considered that, on
a moving platform, the forelegs are committed to the main
loading, while the hind legs are deployed to compensate for
changes in speed and direction.

An animal rarely changes its position while the train
is moving, once the parallel pattern across the truck has been
formed (Sutton et al., 1967). This is reflected in the fact
that bruise trim on equal journeys was 0.89 kg/carcase when
there were four stops versus 1.2 kg/carcase when there were
ten stops (Meischke, 1975). Many halts at sidings are typical
of the schedule of goods trains.

Each animal possesses an individual distance beyond which
the presence of congeners constitute an aggression (Bouissou,
1980). Kilgour (1976) noted that beef bulls will fight if put
closer than $5\frac{1}{2}$ m to one another. When the area per animal is
reduced, as in loaded trucks, the animals cannot avoid each
other and the individual distances are constantly violated,
leading to a considerable increase in the number of aggressive
interactions (Bouissou and Signoret, 1971).

The increase in aggressiveness due to the concentration
of many animals and the body to body contact with animals of
higher rank bring about a true social stress leading in particular to an increase in the weight of the adrenal glands of
dominated animals (Bouissou, 1980).

Sutton et al. (1967) observed that the presence of horns
caused discomfort both to the horned animal and its companions.
If three horned animals stood parallel to each other with their
heads together, the one in the middle showed signs of discomfort
by raising or lowering its head in an attempt to avoid the horns
of the other two. Evasive movement to escape from horns also
took place.

Animals with upright horns do not worry each other nearly
as much as do those with horizontal or downward sloping horns
(Bisschop, 1961). The former can keep their horns from being in

the way by holding their necks horizontally, but the latter must hold their heads very high or very low, i.e. in positions they cannot maintain for any length of time.

Lying down in transit

Cattle often lie down in the trucks while they are moving. In a 9½ h journey by road, Anderson (1973) found 25 animals lying down out of a total of 494. Of these, 20 were lying in a normal 'camping' position and were easily brought to their feet (presumably the remainder were in dangerous positions and difficult to bring to their feet). During rail travel the animals do not usually lie down until about 20 h after the journey has started. When they were given rest periods in kraals en route, they remained standing for 13 - 17 h after reloading before the first one lay down (Sutton et al., 1967; Anderson, 1973). In the course of a 96 h journey without off-loading all the cattle lay down at one point or another (Sutton et al., 1967). Bisschop (1961) considered that cattle became accustomed to the train and so the number of wagons in which cattle were found lying down and the number per wagon increased.

Up to 13 h travelling after the first rest period it was easy to force bullocks to their feet. From this journey time on, more animals were found down and were assisted to their feet with increasing difficulty. After 31½ h travelling, some 50% of the heavier bullocks (350 kg dressed weight) and 25% of the lighter bullocks (270 kg dressed weight) were down in the wagons. At unloading (32½ h), 3 of 494 animals required considerable assistance to bring them to their feet, (Anderson, 1973). The period spent lying down averaged 15 min but could be up to 5 h at a time (Sutton et al., 1967).

A greater percentage of animals lay down in the padded than in the unpadded wagons, usually at the front end of the wagon and always across it. The position of the wagon in the train had no influence on the number of cattle lying down. No correlation was found between the number of animals lying down in a wagon and the magnitude or intensity of the bruises found on their carcases.

Standing animals continued to maintain the parallel pattern across the truck although occasionally the animals nearest the one lying down stood lengthwise in the truck with their heads over it. Standing animals were careful not to trample on recumbent ones (Sutton et al., 1967). However, Meischke (1975) found that the animal with the heavier bruise trim weight was down in the truck during the journey. Trampling would be unlikely if the animals were calm or not thrown off-balance (Sutton et al., 1967). If the recumbent animal blocks the exit door during unloading, the other cattle will unload over it. This animal may often be observed to be shivering clearly after unloading.

Behaviour in stationary trucks

In transport by road, the animals became unsettled when the vehicles were stationary (Anderson, 1973). During rail transport, the standing position was rarely changed until the train came to a halt (Anderson, 1973; Sutton et al., 1967). Then the animals frequently moved to face the direction in which the train was moving (Anderson, 1973) or to haphazard positions (Sutton et al., 1967). However, the transverse position was immediately reassumed when the train started to move (Sutton et al., 1967). In one instance, Bisschop (1961) reported that cattle maintained their fixed standing positions during 45 min spent at a siding.

Air transport

During a 24 h air journey, Hereford cattle browsed at hay but did not drink and for the most part appeared unconcerned during the whole flight (Jackson, 1979). The lying/standing behaviour of the animals was assessed at various times during the flight and, as might be expected, there was an increase in the number of animals seen lying down as the journey went on.

BEHAVIOUR DURING HANDLING, LOADING AND UNLOADING

A major factor influencing animal welfare concerns the reaction of handlers to the animals' behaviour. When animals hesitate the handler may resort to more and more violent techniques to impose his will on the animal. Where animal handling facilities hinder rather than aid animal movement the violent techniques may well become the norm. Animal handling facilities vary greatly; however, certain characteristic patterns of behaviour are displayed by cattle during handling and these will now be considered.

Following behaviour

Cattle can be encouraged to go in a desired direction by having one of their own species acting as a goal (Ewbank, 1968), i.e. they will follow the leader (Hafez et al., 1969; Grandin, 1979a). In loading this tendency can easily be taken advantage of by using a single file narrow race. The next animal must be able to observe the animal in front of it moving down the race (Grandin, 1980c). Thus gates across the race should be constructed so that the animals can see through them. Following behaviour can impede cattle movement if animals in adjacent races observe each other moving in the opposite direction. Fences between these races should obviously be solid. Conversely, if animals in two adjacent races are to move in the same direction, the fence between them should be constructed so the animals can see through it (Grandin, 1980c). Isolated animals will attempt to rejoin members of their species who are visible even to the extent of jumping fences.

Circling behaviour

Cattle will turn to face a person when he enters their pen (Grandin, 1980c). As he moves about the pen, the cattle circle in order to keep him in sight. When cattle are driven from directly behind, they tend to turn back to look at the handler, the opposite of the desired outcome (Grandin, 1979a). The most effective position from which to drive cattle is at a 45 to 60° angle from a line perpendicular to the animal's shoulder (Williams, 1978 cited by Grandin, 1979a). If cattle are in a curved chute then circling behaviour, to keep a handler moving along the inner perimeter, in view, will bring the cattle further up the chute (Grandin, 1979a).

Freedom v Dead-end

Cattle will balk if a race leads to a dead-end, e.g. a truck (Grandin, 1980c). This is a problem in loading. However animals will follow each other down a curved race and be almost in the truck before they realise it. An animal in line will see only the rear end of the animal in front of it moving around the curve (Grandin, 1979a). The curve must not be too sharp, otherwise it will appear as a dead-end causing the cattle to balk (Grandin, 1979b).

For cattle the minimum length of chute to take advantage of following behaviour is 20 ft (Grandin, 1979a). If large numbers of cattle are being handled a second single file chute is recommended. If an animal lies down in one chute the other will remain open. This gives herdsmen enough time to bring the animal to its feet without resorting to violent methods (Grandin, 1979b).

Animals are constantly looking for an avenue of escape and will move towards one. This can be exploited in unloading by using a wide and straight chute to provide a clear unimpeded path to freedom (Grandin, 1979a). Two problems arise if there is a gap to the side between the truck and the unloading chute. The first is that the animals are now presented with a choice. They will delay as they need time to decide which way to go. This alone will slow down throughput putting pressure on handlers. Also, an animal may well attempt to escape to the side (Grandin, 1980b). If this happens, the stress associated with recapture may well result in a hot and exhausted animal which as such should not be slaughtered.

Cattle trucks should load from the end rather than the side (Sutton et al., 1967; Anderson, 1973). Side-loading trucks may confront cattle with the decision of where to turn once inside the truck. Sutton et al. (1967) found that when doors were towards the centre some animals would turn around to face the door making it more difficult for others to enter.

Flight distance

All animals including man maintain a distance between

themselves and other animals. This is a form of spatial territory which the animal carries with him. Invasion of this space leads to retreat in order to maintain the critical distance. If the flight zone is invaded too deeply, the animals may attack the invader or turn back and rush past him (Fraser, 1974; McFarlane, 1976). This flight distance varies according to the tameness of the animals. It is only 5 - 25 ft for feedlot cattle and may be almost absent in tame cattle (Grandin, 1979a). To force an animal to retreat, penetration of the flight zone is sufficient. Retreating from the flight zone results in the animal halting. This makes possible the manipulation of the animal. If there is no avenue of escape behind the retreating animal he will turn back and run past the handler (Grandin, 1979a, 1980c). This attempt to escape will end only if the handler retreats away from the cattle (Williams, 1978 cited by Grandin, 1980c). This contrasts sharply with the more usual reaction of moving closer to cattle which start to turn back (Grandin, 1979a), if not actually rushing up to them (Williams, 1978 cited by Grandin, 1980c). The flight distance can be easily exploited by moving along the inside perimeter of a curved race. The handler moves into the flight zone and the animal's reaction is to move forward as far as possible.

If the flight zone of animals in a race is penetrated and they are unable to escape, they become agitated and start to rear up. Handlers who lean over the side of the race cause this problem (Grandin, 1979a, 1980c).

Fear increases the flight distance. Consequently, by exciting the cattle a handler, without even moving, may place himself inside the flight zone (Grandin, 1979a).

Responses to visual and auditory stimuli

Cattle tend to move towards light (Grandin, 1979a). They will often balk at moving from brightly lit to dimly lit enclosures (Grandin, 1980c). Thus, illuminating the interior of trucks at night will help to overcome this.

Cattle have poor depth perception and as a result they often mistake strong shadows for changes in depth (Grandin, 1979a).

They will often balk or refuse to cross a shadow or drain grate (Grandin, 1980c). Thus, sudden discontinuity in the floor level or texture, e.g. slats, may cause problems (Lynch and Alexander, 1973). Lighting around loading chutes or work areas should be even and diffuse (Grandin, 1979a).

Cattle hearing cannot be compared directly to human hearing. The auditory sensitivity of cattle is greatest at 8 000 hz whereas the human ear is most sensitive at 1 000 to 3 000 hz (Ames, 1974). Thus high-pitched noises such as the cracking of whips can stress cattle (McFarlane, 1976). The use of hydraulically activated machinery may be a source of balking and the degree of balking may depend on the location of the motor (Grandin, 1975, 1978a; Bisschop, 1961). The sound of compressed air makes cattle move away (Webb, 1966). Cattle move away from sounds pulsed two to four times per second (Webb, 1966). Thus machinery operating at this frequency may cause distress.

THE BEHAVIOUR OF CATTLE IN LAIRAGE

The European Convention for the Protection of Animals for Slaughter (1979) states that animals shall not be taken to the place of slaughter unless they can be slaughtered immediately (Article 6.1) and that animals which are not slaughtered immediately shall be lairaged (6.2). In practice a so-called rest period is usually unavoidable in order to ensure the continuity of supply of animals for processing. Given the type of lairage available in slaughterhouses it is doubtful if animals are able to rest properly or to recover from the stress of transport (van Logtestijn and Romme, 1980). Behavioural studies have improved our knowledge of methods of reducing stress during lairage (Ammerdown Group, 1980). Even small changes in design often contribute significantly to the improvement of the welfare of the animals, with reduction in both stress and physical damage.

Mixed groups

Mixing of unfamiliar groups of cattle occurs frequently at abattoirs and has been widely reported (Augustini, 1980;

Grandin, 1978b; Kousgaard, 1980; Poulanne and Aalto, 1980; Tarrant, 1980). The high level of agonistic interactions which occur when the animals first meet can create a large amount of stress which appears to be more severe than moderate feed or water deprivations (Grandin, 1978b; Bouissou, 1980). It has been shown that cattle become stressed if they are sorted and regrouped more than once during a 48 h period (Grandin, 1978b). Mixing is counterproductive as it prevents the animals resting and they frequently become exhausted and are sometimes injured - an undesirable outcome both from the animal welfare and economic point of view (Tarrant, 1980).

The behavioural response to mixed penning is most obvious and intense among young bulls (Table 1). There is usually an immediate and high level of aggressive and mounting activity which gradually abates over a considerable number of hours, as the social hierarchy is established and the animals become exhausted (Table 1).

TABLE 1

BEHAVIOUR OF YOUNG BULLS AFTER MIXING. TWO STALL-FED FRIESIANS, AGED 14 - 15 MONTHS, WERE MIXED WITH AN ESTABLISHED GROUP OF 10 SIMILAR BULLS, WITH WHICH THEY HAD NO CONTACT FOR AT LEAST 2 MONTHS. THIS WAS REPEATED 5 TIMES USING DIFFERENT ANIMALS EACH TIME. THE ACTIVITY OF THE NEWCOMERS WAS OBSERVED PERIODICALLY AND EXPRESSED AS THE MEAN NUMBER OF INTERACTIONS (BUTTS, MOUNTS) PER MINUTE ($\pm$ SEM).

	Time after mixing			Significance of decrease in activity between 30 min and 240 min
	30 min (n = 10)	150 min (n = 8)	240 min (n = 10)	
Butting	0.42 ± 0.113	0.21 ± 0.064	0.13 ± 0.063	P < 0.05
Mounting	0.23 ± 0.052	0.25 ± 0.063	0.17 ± 0.074	N S
Being mounted	0.64 ± 0.115	0.75 ± 0.189	0.65 ± 0.098	N S
Total	1.29 ± 0.129	1.21 ± 0.234	0.95 ± 0.099	P < 0.05

N S = Not statistically significant

n = number of animals observed

Statistically significant correlations were observed between the level of physical activity (butts, mounts/min) and rise in blood creatine kinase activity, body temperature and fall in muscle glycogen content (Tarrant, unpublished results).

Other sex categories undergo similar aggressive activity,
although usually less intense mounting activity compared with
bulls, as a consequence of social regrouping (Grandin, 1978;
Bouissou, 1980).

The shape of pens is an important factor. Cattle prefer
to rest along the perimeter fence, also square pens mean that
cattle are more crowded than in long, narrow pens. If bulls
find they are closer than 5.5 m to one another then fighting
will result, even in formed groups (Kilgour, 1976). Bulls are
very sensitive to disturbance and are greatly stimulated by
seeing other socially active cattle, whether fighting or mounting. Mixing strange bulls can best be carried out towards
evening, immediately before feeding (Kilgour, 1976).

Horned cattle

When animals are left undisturbed there is always more
activity among groups of horned cattle than among dehorned
cattle (Meischke, 1975). Heterogeneous groups were always
midway between the horned and hornless groups in activity.
On closer observation, most of the activity seemed to be of
the avoidance type. No serious degree of malicious horning or
aggressive behaviour was observed (Anderson, 1973; Meischke,
1975; Sutton et al., 1967) but many incidences of accidental
horning and impedence of free movement were noted.

The incidence of bruising in horned groups transported
by rail was observed to be 41.8% versus 20.1% in polled cattle
(McManus and Grieve, 1964). Ramsay et al. (1976) observed that
there was a significantly higher amount of bruised tissue on
the carcases of horned cattle transported by truck to the
slaughterhouse.

One Afrikaner grade steer with horns about 45 cm long and
sloping outwards and downwards, continually did damage to the
sides and tops of the steers next to him and was set upon by them
in retaliation whenever possible. At slaughter his was the
worst burised carcase of all the observed oxen (Bisschop, 1961).
Also if horns are not tipped they will grow long enough to make
it impossible for the animal to pass through a single file chute

without turning its head, a feat which is sometimes beyond naive animals (Anderson, 1973; Grandin, 1980a).

Feeding and resting behaviour

During breaks in long journeys Bisschop (1961) found a consistent pattern of behaviour. Cattle were unloaded into pens containing water and lucerne hay and there followed:

(i)   A rather surprisingly short period of intensive feeding, during which the oxen apparently ingested sufficient hay (about 2 kg) to satisfy their immediate demands.

(ii) A short period of fidgeting, walking about, desultory picking at the remaining food, drinking, possibly urinating or defaecating and finally lying down to rest, ruminate and sleep.

(iii) A period of rest for anything up to an hour.

(iv) A period when the animals again felt hungry and during which they were restless and moved about searching for food.

Sutton et al. (1967) found that under similar conditions cattle showed a marked preference to walk around slowly before settling down to eat and drink. Neither food nor drink were consumed avidly. However, the overall consumption per head was similar to that reported by Bisschop (1961). In long journeys rest periods of greater than 8 h in every 24 h must be allowed for consumption of an adequate maintenance diet (Bisschop, 1961; Sutton et al., 1967).

CONCLUSIONS

Information on the behaviour of cattle during and after transport is limited to a few studies. Most of these concern rail travel. Practically no data are available concerning other means of transport to slaughter. In particular, it could be misleading to generalise from rail to road transport. The present conclusions are drawn under these limitations.

Additional information on behaviour is required, particularly in relation to the most common situations (e.g. short and medium haul road transport).

Behavioural norms and indices of stress need to be clearly identified to allow interpretation of observed behaviour. In relation to animal welfare, the short-term, transient nature of the preslaughter situation must be recognised.

Design of animal handling facilities and training of stockmen should take account of standard behavioural patterns. This would improve ease of movement and consequently animal welfare. This humane approach is not widely adopted at present.

Mixing of unfamiliar groups in enclosed spaces (e.g. trucks) is unacceptable. Where mixing is unavoidable it should occur well before loading to allow formation of a social hierarchy and to reduce the problem of faecal slush. Short journeys or rest periods during long journeys are necessary to avoid enforced lying down as a consequence of exhaustion and also to reduce social stress. Dehorning is an important requirement where cattle are transported in groups.

REFERENCES

Ames, D.R., 1974. Sound stress and meat animals. Proc. International Livestock Environment Symposium. ASAE. SP-0174, p 324.

Ammerdown Group, 1980. The transport and slaughter of farm animals. Report of a Seminar held between April 16 - 18th, 1980 at Ammerdown, Radstock, Nr. Bath, UK.

Anderson, B., 1973. Study on cattle bruising. Queensland Agricultural Journal, 99, 234-240.

Augustini, Chr., 1981. Influence of holding animals before slaughter. In: Proc. CEC Seminar 'The Problem of Dark-Cutting in Beef'. Eds. D.E. Hood and P.V. Tarrant, Brussels, 7 - 8 Oct., 1980.

Bisschop, J.H.R., 1961. Transportation of animals by rail. 1. The behaviour of cattle during transportation by rail. J.S. Afr. vet. med. Ass., 32 (2), 235-268.

Bouissou, M.F., 1981. Behaviour of domestic cattle under modern management techniques. In: Proc. CEC Seminar 'The Problem of Dark-Cutting in Beef' Eds. D.E. Hood and P.V. Tarrant, Brussels 7 - 8 Oct., 1980.

Bouissou, M.F. and Signoret, J.P., 1971. Cattle behaviour under modern management techniques. Farm Build. Ass. J., 15, 25-28.

Buchter, L., 1981. Identification and minimisation of DFD in young bulls in Denmark. In: Proc. CEC Seminar 'The Problem of Dark-Cutting in Beef'. Eds. D.E. Hood and P.V. Tarrant, Brussels, 7 - 8 Oct., 1980.

Dewes, H.F., 1979. Transit-related lameness in a group of Jersey heifers. New Zealand Veterinary Journal, 27: 45.

European Convention for the Protection of Animals for Slaughter, 1979. European Treaty Series No. 102, Strasbourg, 10 May 1979.

Ewbank, R., 1968. The Behaviour of Animals in restraint. Ch. 10 In: Fox, M.W. (ed.) Abnormal behaviour in animals. Saunders, Philadelphia, London, Toronto.

Fraser, A.F., 1974. Farm Animal Behaviour. Baillière Tindall, London.

Grandin, T., 1975. Survey of behavioural and physical events which occur in hydraulic restraining chutes for cattle. MS thesis, Arizona State University, May 1975.

Grandin, T., 1978a. Transportation from the animals point of view. American Soc. Agric. Engineering 78, Paper No. 78-6013.

Grandin, T., 1978b. The effect of social regrouping on the incidence of dark-cutting carcasses in beef steers. Presented at the Annual Meeting of American Society of Animal Science, July 10, 1978.

Grandin, T., 1979a. Understanding animal psychology facilitates handling livestock. Veterinary Medicine/Small Animal Clinician, May 1979, 697-706.

Grandin, T., 1979b. Designing meat packing plant handling facilities for cattle and hogs. Transactions of the ASAE, 1979, 912-917.

Grandin, T., 1980a. Bruises and carcass damage. International Journal for the Study of Animal Problems, 1 (2), 121-137.

Grandin, T., 1980b. Designs and specifications for livestock handling equipment in slaughter plants. International Journal for the Study of Animal Problems 1 (3), 178-200.

Grandin, T., 1980c. Observations of cattle behaviour applied to the design of cattle-handling facilities. Applied Animal Ethology, 6, 19-31.

Hafez, E.S.E., Schein, M.W. and Ewbank, R., 1969. The behaviour of cattle. In: Behaviour of Domestic Animals. E.S.E. Hafez (ed.) BTC London, England.

Jackson, W.T., 1979. Behavioural observations on farm animals in transit. Applied Animal Ethology, 5, 291.

Kousgaard, K., 1981. Development of special feeds for young bulls kept in overnight lairages. In: Proc. CEC Seminar 'The Problem of Dark-Cutting in Beef'. Eds. D.E. Hood and P.V. Tarrant, Brussels, 7 - 8 Oct., 1980.

Kilgour, R., 1976. The behaviour of farmed beef bulls. NZ Journal of Agriculture. 132 (6), 31-33.

Lynch, J.J. and Alexander, G., 1973. Animal behaviour and the pastoral industries. In: The Pastoral Industries of Australia. Eds. G. Alexander and O.B. Williams. Sydney University Press. pp 371-400.

McFarlane, I., 1976. Rationale in the design of housing and handling facilities. In: Beef Cattle Science Handbook, Vol. 13 (M.E. Ensminger ed.). Agraservices Foundation, Clovis, Calif., 223-227.

McManus, D. and Grieve, J.M., 1964. Bruising of cattle stock for slaughter. Veterinary Record, 76 (3), 84-85.

Meischke, H.R.C., 1975. Bruising in cattle. A report to the Australian Meat Board.

Poulanne, E. and Aalto, H., 1981. The incidence of dark-cutting beef in young bulls in Finland. In: Proc. CEC Seminar 'The Problem of Dark-Cutting in Beef'. Eds. D.E. Hood and P.V. Tarrant, Brussels, 7 - 8 Oct., 1980.

Ramsey, W.R., Meischke, H.R.C. and Anderson, B., 1976. The effect of tipping of horns and interruption of journey on bruising in cattle. Australian Veterinary Journal, 52, 285-286.

Sutton, G.D., Fourie, P.D. and Retief, J.S., 1967. The behaviour of cattle in transit by rail. J.S. Afr. vet. med. Ass. 38 (2), 153-156.

Tarrant, P.V., 1981. The occurrence, causes and economic consequences of dark-cutting in beef - a survey of current information. In: Proc. CEC Seminar, 'The Problem of Dark-Cutting in Beef'. Eds. D.E. Hood and P.V. Tarrant, Brussels 7 - 8 Oct., 1980.

Van Logtestijn, J.G. and Romme, A.M.C.S., 1981. Animal welfare in relation
to transport, lairage and slaughter in cattle (a review). In: Proc.
CEC Seminar, 'The Problem of Dark-Cutting in Beef', Eds. D.E. Hood
and P.V. Tarrant, Brussels 7 - 8 Oct., 1980.
Webb, T.F., 1966. Feasibility tests of selected stimuli and devices to
drive livestock. ARS 52 - 11. Transportation and Facilities Research
Division, Agricultural Research Service, USDA.

DISCUSSION

G. von Mickwitz *(Federal Republic of Germany)* Does a resting time for boars
of 1 h or a little longer result in better pH and a better price for the
carcases?
 Whether cattle will lie down or not depends on the loading density, as
has already been seen with the transport of cattle in Africa. What was the
loading density in the trial?

P.V. Tarrant *(Ireland)* In relation to rest periods after transport, I think
everything depends on the nature of the resting environment. If the resting
environment is good, things are not going to get worse, but they will improve
only slowly, especially in the case of cattle. My comment also referred to
the usual absence of adequate resting conditions in slaughterhouses. Under
very good conditions one would anticipate a very gradual improvement and I
suppose that even one hour might make some contribution. I feel, however,
that in relation to the meat quality difference in cattle, one would need a
resting period of 24 h to obtain a substantial improvement.
 The density of loading must be a very important factor in relation to
whether cattle lie down in transit. If they are loaded too loosely you will
have animals falling, and if you have them packed too tightly, there is the
danger that once they have laid down they will not be able to get up again.
I cannot recall the exact loading densities used in the studies by Sutton
et al. and Bisschop.

E. Wagner *(Luxembourg)* Are there any references made to breeds in the
observations made during transport? Normally as an animal breeder you are
told that Limousin are very nervous or that Charolais are very quiet cattle.
Are there any differences between different breeds?

P.V. Tarrant In the studies which we consulted the breeds were usually mixed
and occasionally, in the case of railway transport, Afrikaner cattle were the
ones studied. This has little application for our European breeds. Usually
the breeds are mixed and there has been no careful examination of behaviour
in relation to breed. We confined our studies to the pre-slaughter period,
but perhaps there has been some observation of the differences between dairy
and beef breeds in the production phase.

SESSION III

SIGNIFICANCE OF THESE CHANGES AND EFFECTS
IN RELATION TO HEALTH AND WELL-BEING

Chairman: D.T. Smidt

LOSSES CAUSED BY TRANSPORT OF SLAUGHTER PIGS IN THE NETHERLANDS

J.G. van Logtestijn, A.M.T.C. Romme and G. Eikelenboom

Department of Science of Food of Animal Origin,
Faculty of Veterinary Medicine,
Biltstraat 172, Utrecht, The Netherlands.

INTRODUCTION

Losses related to transport of slaughter pigs can be divided into different aspects.

Animal welfare

A decrease in animal welfare is one of the most important disadvantages associated with the transport of slaughter animals. Transportation means for an animal the loss of familiar environment, exposure to many foreign stimuli, an unusual muscle activity etc. Yet the extent of decrease in animal welfare is often difficult to assess since there are frequently no objective methods.

Animals which die during or after transport and those which produce inferior meat quality (pale, soft and exudative (PSE) and dark, firm and dry (DFD) meat) in their carcase after normal slaughter

These effects, which are more directly associated with economic losses, are easier to determine. Only a small part of the carcases with inferior meat quality is detected by the meat inspection services. The main reason is that the relationship between visual judgement at 45 min *post mortem* and ultimate meat quality is restricted. In general only extreme cases of PSE are detected. Nevertheless part of the meat of these stress-susceptible pigs is declared unfit for human consumption. Three categories of losses related to transport can be distinguished in relation to meat inspection:

a.   Dead animals which are in principle to be rejected.

b.   Animals slaughtered in emergency. The decision as to

approval, conditional approval or rejection depends on the findings of the inspection after slaughter.

c. Pigs slaughtered in the normal way, which in some cases show bruises, skin discolourations, blood splashes (due to stunning), PSE and DFD meat.

Labour aspects should also be mentioned here. The human labour involved in transport of slaughter pigs is in general rather hard and stressful. An improvement of the conditions for pigs, particularly with regard to loading and unloading, will often coincide with an improvement of labour conditions for man, as is the case with the use of the hydraulic lift system, for instance.

DEATH RATE

Up until 1970 the death rate during transport of slaughter pigs in the Netherlands increased to a dramatic average value of $7^{o}/oo$ in some slaughterhouses. In a pig slaughterhouse with a supply of approximately 400 000 pigs per year the death rate, including loss in the pens, developed as shown in Table 1.

TABLE 1

DEVELOPMENT OF DEATH RATE FOR PIGS DUE TO TRANSPORTATION (1960 - 1980)

1960	:	1.5 $^{o}/oo$	1972	:	5.2 $^{o}/oo$
1964	:	2.8 $^{o}/oo$	1976	:	3.8 $^{o}/oo$
1968	:	4.7 $^{o}/oo$	1980	:	2.1 $^{o}/oo$

Since 1965 this problem has been of much concern, due to the fact that it is directly related to the percentage of so-called PSE and DFD meat in the slaughtered pigs.

In 1972, 1976 and 1980 an inventory was made in many slaughter plants, and among other things the monthly death rate during transport and in the pens of the slaughterhouses was registered. This inventory involved about 45% of the total number of pigs slaughtered in the Netherlands in these years. The total death rate, including losses during lairage, decreased from 4.2°/oo in 1972 to 3.7°/oo in 1976 and 3.0°/oo in 1980 (Corstiaensen et al., 1977; De Bruin, 1967; Lendfers, 1974; Minkema et al., 1977; Sybesma et al., 1978).

The number of deaths in the lairage varied considerably between the different slaughterhouses, but also showed a steady decrease. On average it was 0.7°/oo in 1972 and 0.5°/oo in 1980 (Figure 1).

Seasonal fluctuations are a well known phenomenon: transport losses are much higher during the summer months than in the winter months due to a higher temperature and especially a higher water vapour pressure in the air (Figure 2). In 1980 the death rate was also highest during the summer months, although seasonal differences were much lower than in the preceding years (Figure 3). The average death rates registered in 1980 in the different slaughterhouses show striking variations: from 2.0 to 4.7°/oo!

Unfortunately, reliable recent data on the change in the incidence of PSE and DFD meat during these years are not available. The positive association between death losses due to stress and the occurrence of PSE suggests, however, that improvements have been made in this field and this is confirmed by reports from practice.

MEASURES TO DECREASE TRANSPORT LOSSES

There are different possibilities for effective measures against transport losses.

Improvement of the genetic status

Most cases of transport losses can be attributed to the Malignant Hyperthermia Syndrome. This syndrome is genetically determined by an autosomal recessive gene. Stress susceptible piglets can be selected out at an age of 6 - 12 weeks by means of the halothane test. Since 1977 the test has been applied in the Netherlands by herdbook and breeding companies in their breeding programmes. As a result the percentage of reactions in the Dutch Landrace pig has been drastically reduced (Eikelenboom et al., 1978; Minkema et al., 1977).

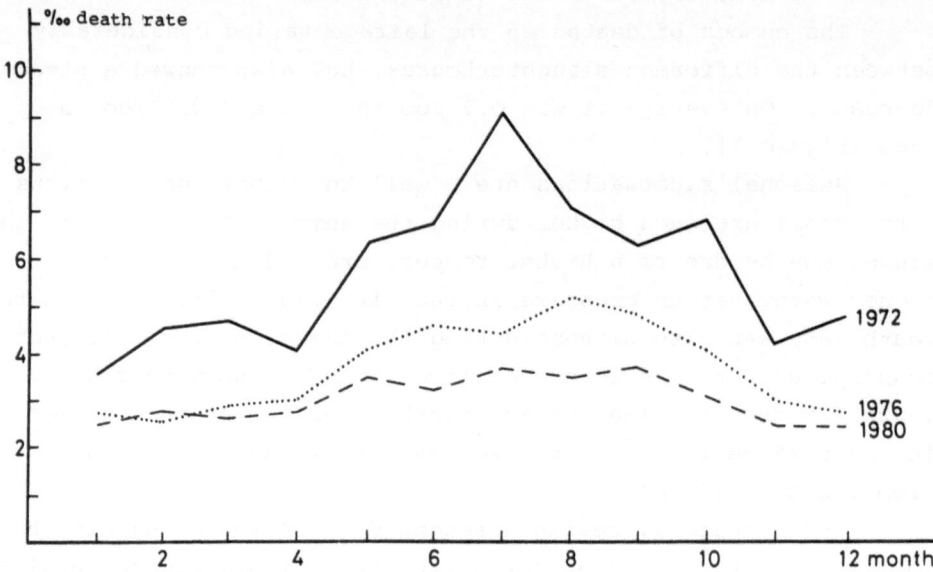

Fig. 1.   Transport losses (TD + DOS) in 1972, 1976 and 1980.

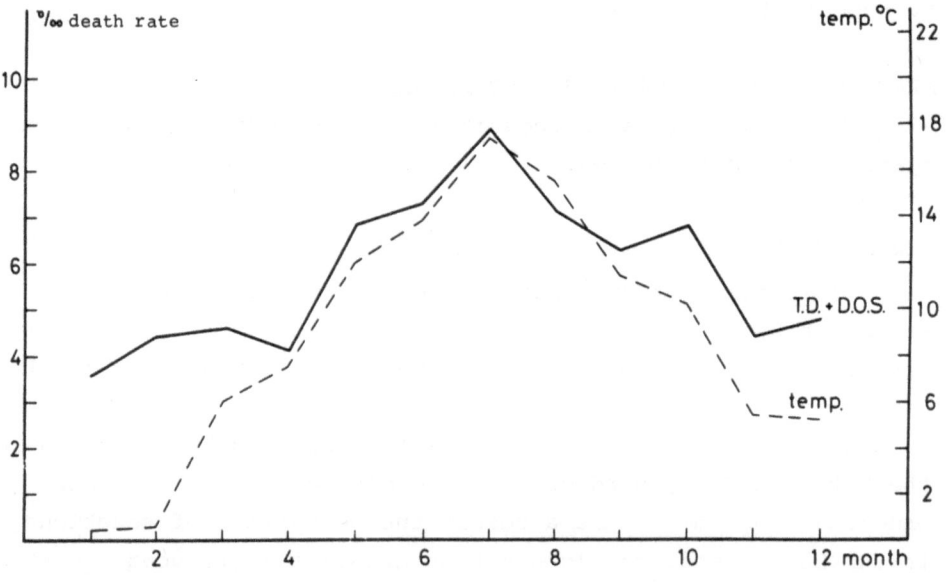

Fig. 2.   Death rate—temperature 1972.

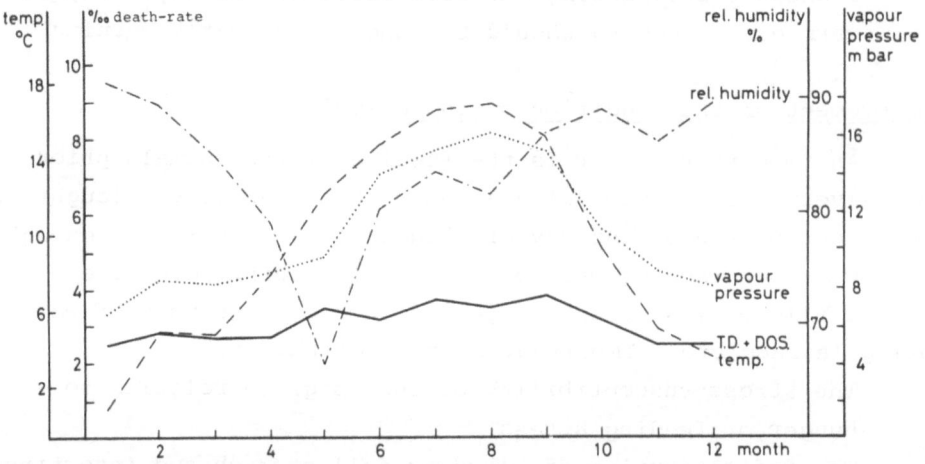

Fig. 3. Death rate⌒temperature 1980.

Information and instruction

Part of the transport losses is related to rough, unmotivated and otherwise inadequate treatment by people who have to handle and transport the pigs. The situation can undoubtedly be improved by continuous and effective information, instruction of these people and legislative measures and inspection.

Improvement of the technical conditions

Much can be done to improve the methods of loading and unloading and particularly the means of transport (Lendfers, 1974; Van Putten and Elshof, 1978; Sybesma et al., 1978):
- The use of loading ramp or dock or special delivery pens in the farms.
- The use of a hydraulic lift and correct use of partitions in the truck, the use of suitable bedding material such as straw, wood shavings and sawdust in the vehicle to avoid slippery floors.
- Optimal ventilation, especially during transport of long duration and during warm weather.
- Avoidance of the use of electrical prods.
- Overcrowding should be avoided.
- Driving at constant speeds with a minimum of severe braking.

- Transport only during the cool parts of the day. So, on warm days, loading should be done in the early morning.

Improvement of the condition of the animals

An important factor is the feeding of the animals prior to delivery. It is generally assumed that pigs to be slaughtered should not be fed on the day of slaughter. However, not enough is known about the optimal conditions of feeding, e.g. the optimal time of last feeding, feed composition, supply of water during lairage etc. Important aspects of this are:
- The stress-susceptibility of the pigs in relation to hunger or feeding stress.
- The redistribution of blood by full stomach and intestines, the generation of heat by ingestion, the impairment of lung function thus reducing the evaporative heat losses.
- It is questionable if water should be freely available.
- The supply of water and food has a distinct influence on the filling of stomach and intestines, on the risk of opening these organs during slaughtering and the consequent risk of spreading enteropathogenic bacteria through the carcase.

Optimal organisation of the delivery of pigs

A good delivery scheme can help to alleviate problems around unloading and time of stay in lairage. The optimum duration of stay is 2 - 4 h. A delivery scheme has to depend on the supply of slaughter pigs, the slaughter capacity, the number and size of pens available for lairage, avoidance of mixing of different groups, climatic conditions and available means of transport.

Priority has to be given to slaughter of pigs which arrive in the slaughterhouse as casualties. In general, the decision that it is impossible to take into account an optimal lairage period for the pigs to be slaughtered is often too rapidly taken. In many cases, however, much more can be done.

Risk of transport losses

Insurance against transport death loss is common. Farmers

who use their own transport lorries in general appear to have
much lower transport death losses in pigs than the normal private
drivers. (Sybesma et al., 1978). Although an insurance against
excessive transport losses is useful, especially for farmers and
drivers operating on a smaller scale, the insurance system should
perhaps be reconsidered. Perhaps a certain division of the risk
between the insurance company, the transport company and the farmer is the most appropriate and honest system, for reducing transport losses more effectively.

Use of drugs

The use of neuroleptics and other drugs may cause residue
problems and may result in a negative image of the product to
the consumer. Furthermore it is questionable if the use of
these drugs is necessary and gives the results in normal
slaughter pigs which many people seem to expect. Moreover, it
is preferable to find a causal and not a symptomatic solution
to the problems. For these reasons, the use of drugs has to
be avoided.

Showering in the pens

Showering of pigs during lairage is widely applied in
the Netherlands, although not much is known about the necessary
conditions, method of administration and amount and temperature
of the water. The animals become quieter, they are cleaner and
electrical stunning is improved etc. Recently we have carried
out a number of experiments on the effects of showering on meat
quality *post mortem* under various environmental temperature
conditions. Our preliminary results obtained so far are
promising.

REFERENCES

Corstiaensen, G.P., De Bruin, J.J.M., Lendfers, L.H.H.M., van Logtestijn,
J.G. and Verdijk, A.Th., 1977. Pig losses during transport and in
the pens of slaughterhouses in the Netherlands in 1976. Tijdschr.
Diergeneesk., 102 13: 811-815.
De Bruin, J.J.M., 1967. Some problems in slaughter pigs. Tijdschr.
Diergeneesk., 92 5: 320-322.

112

Eikelenboom, G., Minkema, D., Van Eldik, P. and Sybesma, W., 1978.
    Production characteristics of Dutch Landrace and Dutch Yorkshire
    pigs as related to their susceptibility for the halothane-induced
    malignant hyperthermia syndrome. Livestock Production Sci. 5, 277-284.
Lendfers, L.H.H.M., 1974. Sensitivity of the Dutch slaughter pig to trans-
    port. Thesis, Utrecht.
Minkema, D., Eikelenboom, G. and Van Eldik, P., 1977. Inheritance of MHS
    susceptibility in pigs. Proc. 3rd Int. Conf. Prod. Disease in Farm
    Animals: 203-205, Pudoc, Wageningen, The Netherlands.
Sybesma, W., Westerink, N.G., Corstiaensen, G.P. and Van Logtestijn, J.G.,
    1978. Effective measures and their consequences on transport of
    pigs in the Netherlands. Proc. 24th Eur. Meet. Meat Res. Workers,
    Kulmbach.
Van Putten, G. and Elshof, W.J., 1978. Observations on the effect of trans-
    port on the well-being and lean quality of pigs. Animal Regulation
    Studies 1, 247-271.
Verdijk, A.Th.M., Van Logtestijn, J.G. and Van de Meij, D., 1974. Management
    of live pigs for slaughter in abattoirs. Tijdschr. Diergeneesk.
    99, 3-9.

DISCUSSION

G. van Putten *(The Netherlands)* Does a broadcast of the expected weather
conditions for the next day help? We have a special weather broadcast for
farmers. Would it help if there was a warning not to transport pigs, for
instance after 8.00 the next day, or would it raise too many difficulties
of transportation? I would like your opinion on this. It is done in the
United States.

J.G. van Logtestijn *(The Netherlands)* It would be rather difficult for the
factories to take weather forecasts into account for good delivery schemes.
However, I think it would be a very good thing if these data were given to
the farmers through radio broadcasts. I believe we have to try everything
to reduce transport problems for man and animals.

D. Smidt *(Federal Republic of Germany)* Do you think the forecast is reliable
enough for the next 12 h?

J.G. van Logtestijn I believe so.

M.E.T. Watts *(UK)* I think that insurance is something which should be removed from your list of risks. Once you start thinking about insurance you
are thinking of a capital sum, the animal loses its status, simply becoming
a value, and if there is a cover, nobody is responsible. There will be no
financial loss and so nobody will bother. There will be very little
stockmanship.

J.G. van Logtestijn I do not know whether other people have any experience
of this. We have one factory, the biggest in Holland, where they divide the
responsibility and give a financial interest to the drivers. They had a
death rate of 0.2% last year, whereas without that factor they had a death
rate of 0.4% or 0.6%. That was where there was no responsibility on the
part of the drivers and all losses were paid by the factory.

M.E.T. Watts We have the same for poultry transport in the United Kingdom.
Where the drivers and loaders are being paid a bonus on the lack of deaths
on arrival at the slaughterhouse, the figures are always much better.

K. Fischer *(Federal Republic of Germany)* Have you some experience of how it is possible to transfer knowledge of optimal handling and transportation to the personnel at the slaughterhouse?

J.G. van Logtestijn Are you saying that there is a relationship between the status of the animals and the reaction of the people in the slaughterhouse?

K. Fischer Is it possible to give our information on optimal handling to the personnel in the slaughterhouse?

J.G. van Logtestijn It is a question of training. It can be done by using pamphlets, by information given by organisations to factories.

E. Kallweit *(Federal Republic of Germany)* I think that these training pro-grammes may not be very successful. There are many people in the slaughter plants who know how the animals have to be handled but after all, you always find people in the holding pens who are not too interested. This is a dirty job. I think it would be hard to train them even though this is the most important part before slaughter for meat quality.

J.G. van Logtestijn Nevertheless, in my opinion, training and information are very important. In general, in Holland, these people have low salaries because many people think it is a low category job. In my opinion these people are perhaps the most important in the entire slaughterhouse, and we should pay them a higher salary. Financial interest is also an important incentive.

G. von Mickwitz *(Federal Republic of Germany)* Your figures showed a decrease in percentage. Was this due to better lorries during the last ten years? I think the factories in Holland have done a lot of work on this. The problem is now that we should have smaller lorries with better ventilation rather than big ones. The large ones have to go from one farm to another. The drivers do not like to separate the animals on larger lorries, as this might cause hand-ling problems.

There is another problem when you have to separate the pigs into the stunning boxes at the slaughterhouse. In plants where the meat is used or sold immediately after slaughter the man in charge of that operation is in the best position to see what should be done. In plants where the animals are slaughter-ed and the meat is not used, then there will be less care. If that is so, then a very good transport system may be invalidated. There must be good handling from the pen to the stunning box.

W. Müller *(Federal Republic of Germany)* Do you think that the higher air velocity produced by fans could have the same effect as a shower?

J.G. van Logtestijn I am not a physiologist, but if the pigs produce abnormal amounts of heat they can try to lose this heat by respiration. Only part of the heat is lost by radiation through the skin. Another factor is the loss of heat by contact, such as contact with cold water. Ventilation is only important in relation to respiration. I believe showering is more important than ventilation.

W. Müller I agree, but there are some dangerous situations as well. If you have a high ambient temperature, perhaps 28°C, and you use a shower, the relative humidity comes to 100%. Reducing heat by respiration is not neces-sary with the shower because there is radiation and loss of heat by contact with the cold water, but if you stop the shower there is 100% relative humid-ity and there is no way of removing water vapour from the respiratory tract into the ambient air. This is a major problem. I would agree that we need showers, but once the showers stop we need fans with an air velocity which will reduce the relative humidity from 100% to 95%.

A. Cuthbertson *(UK)* I am not quite so pessimistic about the role of education in trying to improve attitudes towards animal handling. Over a period of time, through schools, the standards of meatworkers and others could be improved. In the short term, I do not see much happening unless you can demonstrate to the meat industry that there are financial benefits in changing the way they do things. If you are suggesting, for example, that the people who handle the animals before slaughter should be paid more, can you say with honesty that that is going to be worth the added cost to the factory? If you cannot, then it will be difficult to convince the meat industry to change.

The other point I wanted to make was in relation to insurance. In certain countries this leads to shelving of responsibility. In England, for example, insurance of stock is widespread. People feel protected because they know that the animals are insured and so they do not take quite as much care in their work as they might do.

J.G. van Logtestijn I believe that money is important. It is, however, one of many factors, and I believe, as you do, that good training of people and information is very useful. I believe that insurance is a very good thing, but at the same time you have to give responsibility to those who are in charge of the pigs at certain stages of transport.

D. Smidt Could the insurance fees be calculated on the basis of the death losses which the farmer has, as is done with car insurance? This would decrease the negative effect.

C. Platt *(UK)* Have you any figures which relate to the percentage of mortalities which occur actually on the truck compared to those which die after unloading at the slaughterhouse?

J.G. van Logtestijn In 1980 2.5 pigs per thousand died during transport and 0.5 pigs per thousand after transport whilst in the lairage. In 1972 this figure was 0.7 for those which died in the lairage and 6.5 I think for those which died during transport. The number of pigs which die in the lairage has decreased but the number which die during transport has decreased even more.

G. van Putten If I understood you correctly, you guessed that the number of PSE pigs had decreased considerably in recent years to about 5%. Dr. von Mickwitz said that this information from the slaughterhouses is wrong; they investigated and found that a multiple of this percentage is the real figure. I think we have two contradictory statements. What is your opinion?

J.G. van Logtestijn We had to try to distinguish between figures from meat inspection, which only detects extreme cases of PSE and DFD meat, and figures from the slaughterhouse, where all cases are looked at in the dissection room. If you dissect meat you get a good idea of its quality but meat inspection takes place immediately after slaughter and then really only in extreme cases. Thus, I would guess the real figure should be between 5-10% these days for Holland.

G. von Mickwitz The dead animals are the tip of the iceberg, and the iceberg itself is the 20% and 40% PSE. Meat inspection only shows up 0.4%. If you do your own research you will find from 20% up to 30 and 40%. You will definitely find 20%.

N.J. Nielsen *(Denmark)* Could the lowered PSE frequency and the lower death rate be due to the halothane test in your selection programme? I know you have used this for some years, and you should have a lower frequency of halothane sensitive pigs.

J.G. van Logtestijn I believe that the selection on the basis of stress susceptibility is one of the major factors in this phenomenon, as is the use of cross-breeding between Yorkshire and Landrace pigs.

RECENT RESULTS FROM INVESTIGATIONS OF TRANSPORTATION OF PIGS FOR SLAUGHTER

N.J. Nielsen

Danish Meat Research Institute, Roskilde, Denmark.

ABSTRACT

In Denmark a development towards a more standardised transport of pigs must be expected in the coming years. On the basis of our most recent experiments, which were carried out under practical conditions, a series of recommendations have been worked out which involve both producers and hauliers. Producers must prepare the pigs for collection, i.e. slap-mark them, move them to collection pens with a water supply on a previously agreed day, and refrain from feeding them on the day of collection. Pigs from different pens should not be mixed in the collection area.

Hauliers should have well-equipped lorries with effective mechanical ventilation, partitions, automatic loading devices such as a tailgate lift, as well as a special, non-skid, rubber surface on the floor of both the lorry and lift. Loading, transport and off-loading should be carried out with the necessary consideration for the pigs. Loading intensity should be a maximum of 1 pig per 0.35 m² of floor area.

There is a close relationship between a considerate treatment and meat quality and mortality during transport and lairage. A considerate treatment is ensured if at the same time lorry personnel and other people involved are given adequate instruction. It is absolutely imperative that all personnel have the necessary knowledge, and possess those personal qualities, which - together with a well organised and well co-ordinated transport and lairage - ensure the desired stability and a uniform, considerate treatment of slaughter pigs.

INTRODUCTION

During the last 20 years the structure of slaughter pig farms in Denmark has changed a great deal. In 1960 about 5% of slaughter pigs were supplied from farms with an annual production of more than 500 pigs. Today, over 50% of slaughter pigs come from large farms. At the same time the number of farms has fallen from just over 150 000 to 70 000.

It is clear that this change in structure opens up possibilities for changing methods of delivery and transport. However, there are still many small farms who produce irregularly and often only 1 - 2 pigs per week. One of the reasons for this is that most pigs are supplied within an extremely narrow weight range, around 90 kg liveweight.

In Denmark most producers are members of the local co-operative bacon factory and send their pigs for slaughter there. Thus, delivery is always direct from the producer to the factory and transport distances are fairly short, only rarely exceeding 100 km.

Even though some Danish bacon factories are in the process of making changes, there are still differences between factories with respect to the treatment pigs receive. These differences must be eliminated in the long term. A large experiment has therefore been carried out on one Danish bacon factory to investigate the way in which collection, transport and lairage procedures as a whole should be carried out to attain various goals. Among other things the investigation meant that the factory's seven transport lorries were re-equipped and 250 000 pigs were investigated for meat quality.

OBJECTIVES FOR DELIVERY AND TRANSPORT

All arrangements for collection and transport should be based on ensuring the following:
- a considerate treatment
- a good, uniform meat quality
- a low mortality during transport
- a delivery which protects the health status of the herd
- a rational collection and transport

Certain guidelines for the transport of pigs are given in Danish transport regulations (Government notice on transport of animals No. 208, 17th June 1964). Among other things, it is stated that pigs should have sufficient room, adequate ventilation and protection against the elements, e.g. wind, cold and rain.

Thus, a floor space of at least 0.35 m^2 per pig is required and no more than 30 pigs are allowed in any one partitioned area of the lorry. Although these are important factors, it is just as important that the lorry driver and the producer have those personal qualities that are all-important for a considerate treatment during removal from the pens, transport to the factory and off-loading there.

On the basis of both the recently concluded experiment and the objectives for delivery and transport, the Danish Meat Research Institute has proposed a more detailed programme for delivery, transport and lairage under Danish conditions. In this programme the producer, the lorry driver and the factory must closely co-ordinate their arrangements so that improvements in one area are not negated by unfortunate arrangements in another. The treatment of a slaughter pig from the producer up to the point of stunning should be considered as a whole and must be harmonised throughout. There should be a high degree of co-ordination and control.

The programme is based on the fact that there is a close relationship between the treatment pigs receive and their meat quality. All Danish investigations carried out in recent years have shown that meat quality is influenced among other things by energy reserves (glycogen) in the muscles at the point of slaughter. As shown in Figure 1, all exhausted pigs with a low energy content at sticking will develop DFD meat, while all pigs with a heritable disposition for PSE meat will develop PSE meat, when the energy content in muscles is high, (DFD = Dark, Firm and Dry, PSE = Pale, Soft and Exudative).

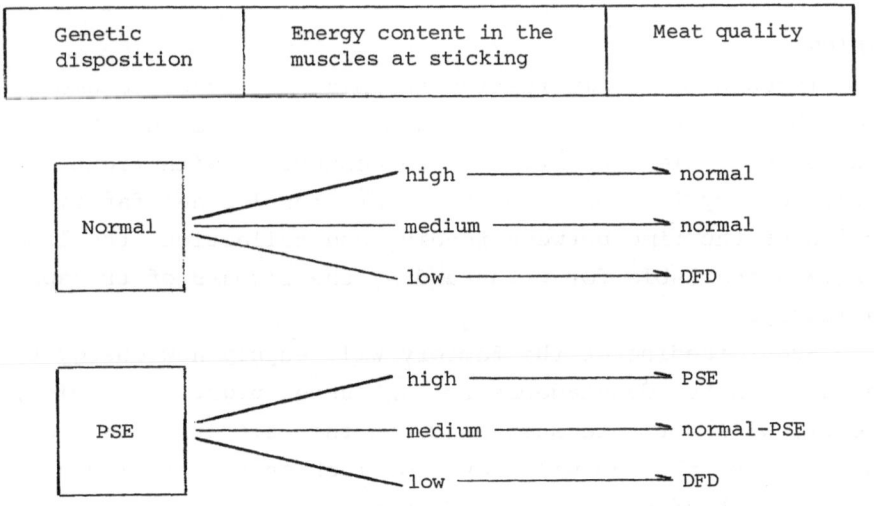

Genetic disposition	Energy content in the muscles at sticking	Meat quality

Fig. 1.

Various factors in the pre-slaughter treatment are important for the energy content in muscles. These factors express to a certain extent how considerate the treatment has been before slaughter. Figure 2 shows those factors of primary importance for the energy content of muscles.

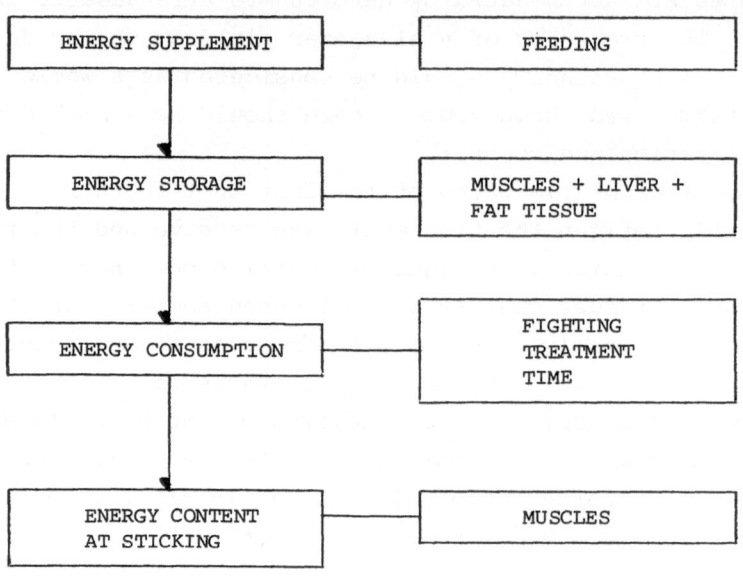

Fig. 2. Factors of primary importance for the energy content of muscles.

FEEDING

Through the daily feeding the pigs are supplied among other things with energy for growth and maintenance. The energy is used in muscle activity e.g. in connection with fighting. Surplus energy is stored in the liver, muscles and fat tissue. The longer the time between feeding and collection, the less energy is available for withstanding the rigours of transport and lairage.

Sugar feeding at the factory will supply new energy to pigs and can be advantageous for exhausted pigs. Some pigs, however, will not take sustenance in this situation and moreover, many of those that do will have too much energy at slaughter with the risk of developing PSE meat.

TRANSPORT AND LAIRAGE PROCEDURES

During transport and lairage the energy consumption of pigs will be greatly dependent on the treatment they receive. Energy consumption will be quite different from pig to pig. Some pigs will have a low consumption, while others, especially if they fight, will have a high energy consumption. Improving transport lorries and holding conditions will lead to certain reduction in energy consumption, but as long as fighting takes place, there will always be some pigs which consume too much energy.

TRANSPORT AND HOLDING TIME

It goes without saying that the duration of the transport and lairage is extremely important for the amount of energy pigs consume; the longer the time, the greater the energy consumption. However, when considering the time factor, it must be borne in mind that pigs will begin to mobilise energy from depots in the liver and fat after a certain time and thus will rebuild the energy content of muscles.

THE TREATMENT'S 'BASIC LAWS'

On the basis of the relationship between treatment, energy level and meat quality the following 'Basic Laws' can be given.

- the incidence of DFD meat is limited by giving the pigs as considerate a treatment as possible with as little fighting as possible, thereby preventing exhaustion;
- the incidence of PSE meat is limited by giving the pigs a suitable length of time from the last feeding to slaughter, thereby obtaining a suitably low level of energy in the muscles.

If both these conditions are fulfilled, a large proportion of the pigs will reach slaughter with a suitably low energy level in the muscles. All pigs will have a suitably low level of energy and only a few will have consumed too much. The incidence of DFD meat can thus, to a certain extent, express the way in which the treatment has been carried out.

The incidence of PSE and DFD meat can thus give some indication under Danish conditions of how considerate the treatment has been - the longer and more stressful the treatment, the more pigs with DFD meat and the fewer pigs with PSE meat. This was shown by an investigation carried out at the Meat Trade School in Roskilde in 1977. Pigs were kept in special pens in the lairage area for a week to allow them to become accustomed to conditions there. On the day of slaughter half of them were driven about 20 m directly to stunning without any means of force being used, while half were transported and held in the lairage as normal. The results obtained are shown in Table 1.

TABLE 1

| Treatment | No. of pigs | % PSE * | | % DFD ** |
		Gluteus medius	Biceps femoris	Semispinalis capitis, Longissimus dorsi and Semimembranosus
Without transport and lairage	84	33.3	21.4	O
With transport and lairage	83	8.4	4.8	1.2

*   The subjective evaluation of colour and structure used in comparative experiments in Denmark.

** pH value in at least one muscle > 6.5 or pH value in two or three muscles > 6.1 approximately the day after slaughter.

An extremely considerate treatment on the day of slaughter has thus resulted in a higher incidence of PSE meat. Transport and lairage ensure a certain amount of energy consumption and thus fewer pigs with PSE meat. If the transport is also carried out considerately with only a little fighting taking place the incidence of DFD meat will also be low.

These facts were confirmed in every way in the large investigation previously mentioned. Among other things can be mentioned that darkening the lairage caused the pigs to relax

more quickly - but only if they had enough room. If there
was not sufficient room, the pigs did not lie down even after
several hours of lairage, probably because they could not find
a position in which they felt safe. They just crowded together
tightly packed. This affected the number of exhausted pigs
with DFD meat, as shown in Table 2.

TABLE 2

Treatment	m^2 per pig	No. of pigs	% PSE * *Gluteus medius*	% DFD ** *Semispinalis capitis, Longissimus dorsi*
Dark lairage	0.5	860	13.1	18.2
	0.8	318	13.5	4.6
Light lairage	0.8	321	13.4	11.8

*   The subjective evaluation of colour and structure used in comparative
    experiments in Denmark.

** pH value in at least one muscle > 6.5 or pH value in two muscles > 6.1
    approximately the day after slaughter.

In conclusion, various investigations have shown that
factories can improve meat quality considerably via a controlled
and more considerate treatment of slaughter pigs. At the same
time losses during both transport and lairage can be reduced.
As shown in Table 3 an effective mechanical ventilation in
transport lorries can also lead to a reduction in losses in the
lairage.

TABLE 3

Ventilation	No. of pigs	o/oo losses		
		Transport	Lairage	Total
With	119 373	0.24	0.68	0.92
Without	218 416	0.46	1.06	1.52

INSTRUCTIONS FOR DELIVERY AND TRANSPORT

In practice the producer's most important tasks in
connection with delivery are as follows:

to notify the number of pigs correctly, so that the
haulier and factory can organise the transport and
slaughter.

to slap-mark the pigs with the producer's number. The
marking can be carried out using a dye up to 3 weeks
before delivery e.g. in connection with weighing. In
this way long loading times and rough treatment of the
pigs are prevented. Pigs fight most when the lorry is
stationary.

to refrain from feeding on the day of collection. Feeding
increases the incidence of PSE meat and when it occurs
just before collection it increases the number of transport deaths too.

to ensure that pigs have access to water in the collection
pens.

to place the pigs in the immediate neighbourhood of the
loading area, so that the loading times are reduced to
the shortest possible. This protects the health status
of the herd and reduces fighting and mortality among pigs
already on the lorry. The only possible solution is to
have collection pens, when the fact that pigs are not to
be fed is taken into consideration. Pigs from different
pens should not be mixed to prevent fighting. There
should be access to water and it should be possible, if
delays in collection occur, to feed the pigs. It is
important that collection takes place on a certain day
and that postponements only occur rarely. This should
be possible if the transport is correctly organised, and
if producers are accurate in notifying the pigs ready for
slaughter.

to ensure good access to the loading area, both in summer
and winter. A rational transport demands good access for
the lorry and that there is sufficient room for it to turn.

The lorry driver's most important tasks are:

<u>to have a well-equipped lorry</u> with a covered deck, effective mechanical ventilation and ventilation openings both low down and high up on the sides, as well as a tailgate lift, partitions and a non-skid rubber surface on the floor.

<u>to have a clean lorry</u> before collecting the first pig.

<u>to spread sawdust on the floor of the lorry</u>. Rubber mats lower the amount of sawdust necessary.

<u>to load pigs considerately</u>, and close off a partitioned area as soon as it is full.

<u>to refrain from overloading</u>, i.e. a maximum of 1 pig per 0.35 m^2 floor area.

<u>to take rest breaks without pigs in the lorry</u>, so that the shortest possible transport times are attained.

<u>to drive without sudden breaking or violent turns</u>.

<u>to off-load pigs considerately</u> by giving the first pigs sufficient time to walk off the lorry by themselves, then driving the rest off the lorry using a push-board in such a way that the group is kept together. The use of electric goads is not necessary.

INSTRUCTION AND FOLLOW UP

In Denmark a development is apparently occurring in the direction of the implementation of all the above-mentioned conditions for delivery and transport. A greater degree of control and co-ordination as well as an improvement in facilities is therefore to be expected. All of this will mean an improvement in the well-being of pigs. However, if this development is to give optimum results, it should be followed up by instruction. A thorough course of instruction should be introduced for all lorry drivers and factory workers handling live pigs. The instruction should include aspects of the treatment and behaviour of pigs, such as the way pigs express suffering. Well co-ordinated and well organised delivery and transport cannot by themselves ensure the required treatment. On the contrary well instructed, suitable personnel can ensure a good treatment, even when conditions are not optimum.

Both the will and the aptitude to treat pigs considerately must be present. Personnel to be used in handling live
pigs should therefore be chosen with great care, and they should
take part in the necessary courses of instruction. The routine,
assembly line nature of the pre-slaughter treatment of pigs
today makes a high degree of control and co-ordination essential.

DISCUSSION

D. Lister *(UK)* I think there is a little confusion here between the effects
of good handling of animals and the effects of low concentrations of energy
substrate in muscle. It is commonly noticed in the UK that the ultimate pH
of commercially slaughtered pigs is always higher than the ultimate pH of the
meat that you find in the pigs slaughtered in our own slaughterhouse, for
instance. We think that this is entirely due to the fact that during commercial handling, Large White cross pigs tend to use glycogen in muscle. This
means that that glycogen is not available to be converted to lactate in the
post mortem period.
    You can do exactly the same thing by restricting the amount of food
given to animals before slaughter and so long as you ensure that there is a
delay of at least 16 h between the last meal and slaughter you will obtain
this effect with great ease. I think, however, that this has nothing at all
to do with the welfare considerations of the animal. These are things which
influence the rate at which the muscle acidifies, and not the extent of the
acidification. We have to be rather careful in concluding that because we
have altered the ability of muscle to become PSE we have brought this about
by good handling. This is simply a reflection of the small amount of glycogen present in the muscle at death.

N.J. Nielsen *(Denmark)* It is clear that commercial handling of pigs gives
differences in ultimate pH. What I meant was that if you handle the pigs for
a long time, you should have a higher frequency of DFD meat because the pigs
will be exhausted. We agree on this. If you change the manner of handling,
making it more considerate and gentle, then more of those pigs which have a
genetic disposition for PSE meat should develop PSE meat. This is why I think
there is a connection between the handling of pigs, their welfare and meat
quality.

D. Lister I do not want to pursue this very far, because I do not think it
is very important, but I think that the effect on the incidence of DFD is due
rather more to nutrition than the handling procedure.

N.J. Nielsen I do not agree, because the amount of energy which has been
used in the hours just before slaughter is a factor. When food was taken
from the pigs 6 h before slaughter, they did not show DFD.

PHYSIOLOGICAL REACTION OF SLAUGHTER ANIMALS DURING TRANSPORT

Chr. Augustini and K. Fischer

**Federal Centre for Meat Research,
Institute for Meat Production and Marketing,
D-8650 Kulmbach, Federal Republic of Germany.**

ABSTRACT

The easily measured parameters of heart frequency and body temperature were tested as possible indicators of stress intensity in pigs. For the basic research, the heart rate and body temperature values of pigs during individual transport were examined. The highest heart frequency peaks appeared when the pigs were loaded and later when driven into the slaughterhouse. After the first minutes in the transport vehicle, heart rates decreased (130 beats/min) to the relatively constant transport value. Rectal temperatures increased during transport to about 0.7°C above the temperature recorded shortly before loading.

During loading by means of a hydraulic tail-gate lift, measured heart frequencies averaged 200 beats/min; during loading with a ramp, heart rates averaged 218 beats/min. Loading by means of the tail-gate lift is less stressful.

When limits of loading density, temperature and relative humidity are exceeded during transport especial stress is placed on the pig. The testing of various combinations of these factors during constant duration of transport (100 min) showed the strong influence of transport distance. In all situations the heart rate decreased with increasing duration of transport. The degree of decrease was influenced by respective treatment combinations.

Without considering relative humidity and loading density, transport at 19°C produced more favourable values than that at 2°C or 29°C. Relative humidity (60 and 90%) and loading density (0.35 and 0.70 m²/100 kg liveweight) led to differing results in each temperature range. The values were obviously strongly influenced by varying levels of physical activity during thermoregulation and by the amount of room available.

The values at loading and at the end of transport were closely correlated. In many cases there was also a highly significant correlation between rectal temperature and heart frequency. The measurements taken before the onset of the imposed stress showed no relationship to those values recorded later.

A connection between the tested stress combinations and meat quality showed a tendency which was just recognisable. On the other hand, significant individual differences were found in animals which were not screened for stress susceptibility. In an experiment with animals pre-tested for halothane positive or negative reaction, it was shown under uniform transport conditions that the significant differences between animals were explained almost exclusively by halothane reaction. A further reason for the difference within treatments was due to differing degrees of muscling.

INTRODUCTION

Transport is a stress for slaughter animals. It can be
supposed that the physiological reaction of the animal provides
the clearest evidence of the stress of transport. Therefore
we examined whether or not the easily measured parameters of
heart rate and body temperature would permit an evaluation of
various stress conditions.

The following questions were posed:

1   Which physiological changes take place during individual
     transport?
2   What is the effect of ramp-loading?
3   What is the influence of transport distance, loading
     density, temperature and humidity?
4   What is the correlation between the reaction to transport
     stress and meat quality?

PHYSIOLOGICAL CHANGES DURING INDIVIDUAL TRANSPORT

Slaughter animals are normally transported in groups.
For gathering basic information, a record of heart frequency
and rectal temperature of a single animal during transport was
made. As research material we used 77 individually raised,
German Landrace barrows weighing 100 kg and the ambient temperature was about 7.0° C. After a 12 h fast the animals were
individually weighed and loaded into the transport vehicle with
a hydraulic tail-gate lift. The animals were transported for
5.3 km which took 14.5 min. At the end of the transport period
the animals were again weighed. Figure 1 records the heart
rate of the 77 singly transported pigs. The highest heart
frequency rates were observed before and after transport. The
values climbed steeply during driving to the weighing point,
they dropped slightly while weighing and climbed even more
during driving to the transport vehicle. Within the first minutes inside the transport vehicle the heart rate dropped sharply.
During transport the rate remained at a relatively constant
level of about 130 beats/min.

During transport no animal lay down but, rather, intensively
investigated the new surroundings. While unloading and weighing
at the end of the transport, the heart rate values again increased

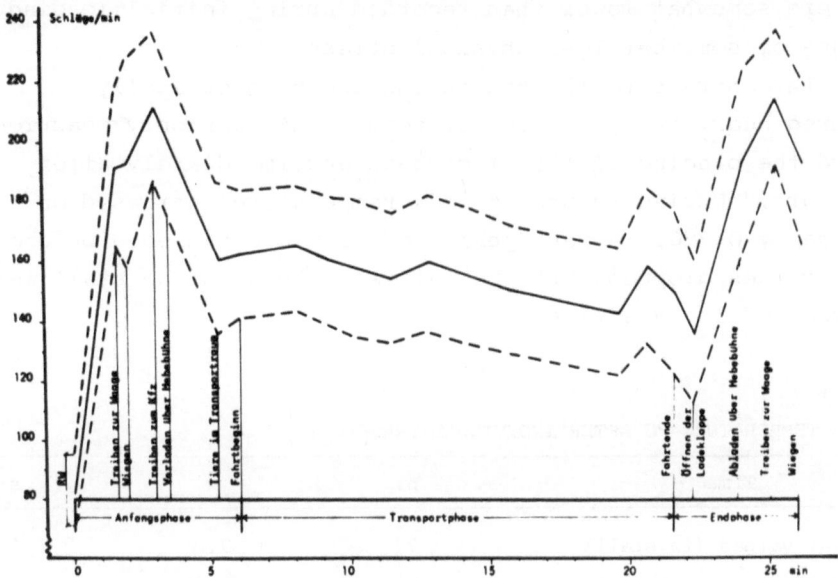

Fig. 1. Heart rate during transport.

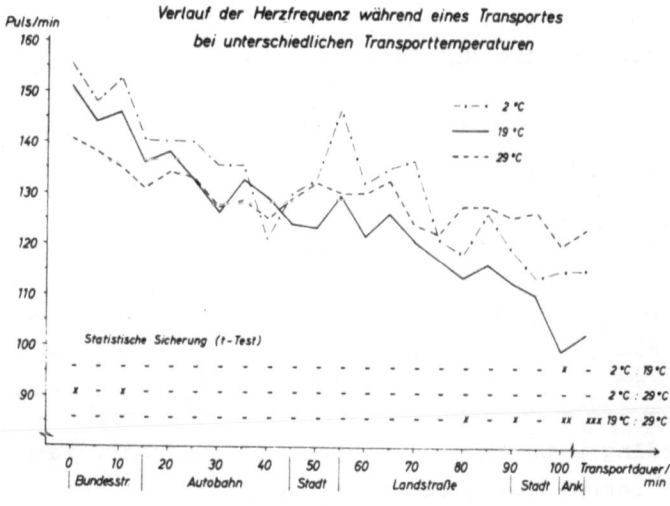

Fig. 2. Heart rate during transport at differing temperatures.

but were somewhat lower than recorded during initial loading
because of somewhat less physical stress.

In contrast to the continuous and telemetrically
measured heart rate, the rectal temperature was only measured
before the placing of the electrodes and immediately after
transport. During transport body temperature increased on
average by 0.7°C. A rest period of 45 min was needed before
these values dropped, but the initial values in the stall were
not reached in this time.

TABLE 1

RECTAL TEMPERATURE $^\circ$C AFTER INDIVIDUAL TRANSPORT

Time	n	$\bar{x}$	s
Initial values (in stall)	77	39.05	0.20
Immediately after transport	77	39.70	0.24
45 min after transport	39	39.48	0.23

EFFECT OF RAMP LOADING

During loading and unloading with the tail-gate lift the
heart frequency dropped. We examined therefore the effect of
ramp loading on heart rate. Several animals were loaded
together. The results are shown in Table 2.

TABLE 2

HEART RATE (BEATS/MIN) DURING RAMP LOADING

	Initial values (in stall)	To transport vehicle	Loading
n	20	20	20
$\bar{x}$	83.1	194.7	217.8
s	12.1	26.7	25.0

During loading with the tail-gate lift the heart rate
dropped whereas ramp-loading led to a steep rise in heart rate.
These values averaged 20 beats/min above the values recorded
while driving the animals to the transport vehicle.

THE INFLUENCE OF TRANSPORT DISTANCE, LOADING DENSITY, TEMPERATURE AND RELATIVE HUMIDITY

Transport stress involves many different factors; we investigated the influence of temperature, relative humidity and loading density during uniform and constant transport (distance and time: 85 km, 100 min). There were 12 different combinations of variables with 2 repetitions of each. The experimental design is shown in Table 3.

TABLE 3

EXPERIMENTAL DESIGN

Temperature °C	Relative humidity %	Loading density m^2/100 kg	Animals/ transport n
2	60	0.35	10
		0.70	5
	90	0.35	10
		0.70	5
19	60	0.35	10
		0.70	5
	90	0.35	10
		0.70	5
29	60	0.35	10
		0.70	5
	90	0.35	10
		0.70	5

It is clearly evident that during transport heart frequency values dropped. The intensity of the decline in heart rate, while primarily dependent on duration of transport, was also influenced by the individual factors of temperature, humidity and loading density.

During the first 40 min of transport a similar drop in values was observed for all combinations of variables: after this time the effect of the imposed stress factors became evident. During transport at 19°C the decline in heart rates continued; at 2°C and especially at 29°C this decline was slower. On arrival at the slaughterhouse these values differed

significantly from the values recorded for transport at 19°C
(Figure 2).

An effect of the respective stress combinations first
becomes recognisable after a relatively long transport time
of 60 - 90 min. The most favourable heart frequency values
were registered with 19°C and 60% humidity. The loading density did not have an effect on rectal temperature even when
taking into account the three differing temperature levels.
We supposed that with a lower loading density of 0.70 m²/100 kg
the animal could maintain effective thermoregulation. However,
at the same time their physical activity increased due to more
available space and it is supposed that here these influences
superimposed themselves. The influence of humidity was different at each temperature level. Higher relative humidity at
2°C resulted in higher heart rates. With this combination we
frequently observed shivering. At 19°C as well as 2°C more
favourable results were recorded with humidity values of 60%.
With 90% relative humidity at 19°C the pigs showed difficulties
with heat dissipation. Pigs that were stressed on loading only
gradually became quiet since they could not maintain effective
thermoregulation. Generally, higher heart frequencies were
found with transport at 29°C but no difference in reactions
were noted between relative humidities of 60 and 90%. Although
many animals were exhausted and often panted on arrival at the
slaughterhouse when both transport temperature and humidity
were high, the heart rates were not higher than at 2°C.
Obviously this subjectively determined state of exhaustion
did not always manifest itself by a clear increase in heart
rate (Table 4).

As a result of the stress during loading rectal temperature climbed on the average 0.5°C above initial values.
However, on arrival at the slaughterhouse, rectal temperatures
had dropped under the initial values. The rectal temperature
in the stall averaged about 0.4°C above the values at the end
of transport. A possible reason for this reduction in temperature could be overcompensation of the body to the higher
temperatures produced during loading. As was the case with
heart rate, rectal temperature with transport at 19°C and

TABLE 4

HEART RATE VALUES IN RELATION TO DURATION OF TRANSPORT, TEMPERATURE AND
RELATIVE HUMIDITY IN THE TRANSPORT VEHICLE

Duration of transport		60 - 90 min		Arrival	
Transport temperature ^{O}C	Relative humidity %	$\bar{x}$	s	$\bar{x}$	s
2	60	122.3	17.07	107.6	17.52
	90	138.1	29.90	122.3	16.80
19	60	108.8	12.18	91.9	14.29
	90	123.6	26.72	113.8	21.11
29	60	130.2	14.05	128.2	20.77
	90	123.2	18.37	118.4	13.05

60% relative humidity was more favourable. Taking all treatments at 29°C together, rectal temperature here averaged 0.4^{O}C
higher than at 2^{O}C and 19°C. These differences were highly
significant. Within the 29°C transport treatments, those with
a high loading density had the highest rectal temperatures.
It appears that in this situation the animals could not dissipate
enough heat even though they panted intensively.

TABLE 5

RECTAL TEMPERATURE ^{O}C IN STALL, AFTER LOADING AND ON ARRIVAL AT SLAUGHTER-
HOUSE AFTER DIFFERING TRANSPORT TEMPERATURES

Transport temperature	Stall		After loading		On arrival	
	$\bar{x}$	s	$\bar{x}$	s	$\bar{x}$	s
2	39.27	0.45	39.08	0.58	38.83	0.61
19	39.39	0.31	39.91	0.35	38.80	0.59
29	39.43	0.37	39.75	0.60	39.22	0.61

Loading had a strong influence on heart rate as well as
rectal temperature. Later values strongly reflected the stress
of loading even after transport for 100 min. The correlation
coefficients between loading and slaughterhouse values were
higher for rectal temperature than for heart rate. The correlation between loading and arrival values was r = +0.22[+] for
heart rate, and r = 0.38[+++] for rectal temperature. The initial
values showed no significant correlation to transport values.

TABLE 6

CORRELATION OF HEART RATES (BEATS/MIN) AND BODY TEMPERATURES ($^{\circ}$C) AT DIFFERENT TIMES IN THE EXPERIMENT

		Heart rate r =	Rectal temperature r =
Initial values	: Loading	+0.29^{+}	+0.17
	: Transport average value	+0.07	-
	: On arrival	+0.07	+0.10
Loading	: Transport average	+0.36^{+++}	-
	: 60 to 90 min after start of transport	+0.32^{+++}	-
	: On arrival	+0.22^{+}	+0.38^{+++}

There was no correlation between the initial values for heart rate and rectal temperature (r = nearly 0); from the beginning of transport, however, highly significant correlations between the two parameters were ascertained.

TABLE 7

CORRELATION BETWEEN HEART RATE (BEATS/MIN) AND BODY TEMPERATURE ($^{\circ}$C)

Rectal temperature	Heart rate				
	Initial values (stall)	Loading	60-90 min after start of transport	Transport average values	On arrival
Initial values	+0.05	+0.05	+0.09	+0.14	+0.10
Loading	+0.07	+0.37^{+++}	+0.41^{+++}	+0.47^{+++}	+0.25^{+}
On arrival	+0.12	+0.24^{+}	+0.50^{+++}	+0.44^{+++}	+0.52^{+++}

CORRELATION BETWEEN THE TRANSPORT STRESS AND MEAT QUALITY

The literature frequently mentions a connection between stress and meat quality. We investigated therefore, whether meat quality is affected by differing transport conditions. Chemical/physical parameters of meat quality (pH value, Göfo value, waterholding capacity) and the concentration of some essential metabolites (glycogen, lactate, ATP) at 45 min *post mortem* were recorded. This investigation showed a barely significant statistical difference between the variables. Only

with the +2°C transport treatment was this relatively homo-
geneous picture different. After transport at 2°C and high
relative humidity significantly lower levels of ATP and glycogen were present. Apparently these pigs had metabolised more
energy than those in the other transport treatments. The 19°C
transport group showed the best meat quality.

TABLE 8

SOME MEAT QUALITY PARAMETERS (M. *LONGISSIMUS DORSI*, 45 MIN *POST MORTEM*) IN
RELATION TO TRANSPORT TEMPERATURE WITH CONSTANT RELATIVE HUMIDITY (90%)
AND LOADING DENSITY (0.35 M^2/100 kg LIVEWEIGHT)

Temperature °C	2		19		29	
	$\bar{x}$	s	$\bar{x}$	s	$\bar{x}$	s
pH-value	5.63	0.30	5.98	0.37	5.79	0.30
Göfo-value	67.9	12.7	77.1	14.2	70.9	16.5
WHC (Grau/Hamm)	6.03	1.80	4.85	2.38	5.21	1.81
Lactate µMol/g	104.37	28.80	84.87	33.09	90.71	21.20
Glycogen µMol/g	8.58	10.86	18.01	16.71	15.06	10.44
ATP µMol/g	1.08	1.50	2.51	2.08	1.10	1.10

The incidence of carcases with a pH value below 5.8 was
quite high. In the M. *Longissimus dorsi* the number averaged
about 53% and in the M. *Semimembranosus* 30.9%. Here again we
could find no clear trend between treatments. We observed
high individual differences within identical treatments. Since
the test animals were not screened for stress susceptibility it
is quite probable that large differences in the genetic disposition within the group made a contribution to poor meat quality.
Correlations of heart rate and rectal temperature to meat quality parameters were low. Of all correlations between heart
rate and meat quality, only those values reported at the end of
loading and on arrival at the slaughterhouse were significant.
Rectal temperatures taken at the slaughterhouse, in contrast
to values measured at other times, were relatively highly
correlated with meat quality. These values were more highly
correlated with M. *Semimembranosus* than those with M. *Longissimus
dorsi*. The considerable individual differences definitely make
such investigations as just described problematic. The genetic
disposition is often the main reason for differences within

and between treatments. In a later trial, therefore, the pigs
were tested with halothane before experimental treatment began.
Similar and standardised treatments of the animal including
transport led to considerable differences between animals which
could be explained to a large extent by different halothane
reactions (Table 9).

TABLE 9

PHYSIOLOGICAL REACTION AND MEAT QUALITY (M. *SEMIMEMBRANOSUS*) AFTER STANDARD-
ISED TRANSPORT STRESS (45 MIN, 18°C, 60%, 0.5 m^2/100 kg LIVEWEIGHT) IN
RELATION TO HALOTHANE REACTION

Halothane reaction	positive			negative			t-Test
	n	$\bar{x}$	s	n	$\bar{x}$	s	
Heart rate (beats/min)							
before loading	71	88.5	14.71	65	83.3	15.90	n.s.
after loading		136.4	17.70		126.9	15.72	++
11 min after start		122.9	14.23		108.6	13.58	+++
21 min after start		132.2	19.60		112.2	17.32	+++
at the end of transport		118.5	18.66		103.8	19.95	+++
Rectal temperature °C at the end of transport		39.12	0.42		38.75	0.53	+++
pH_1-value		5.67	0.28		6.17	0.37	+++
WHC (Grau/Hamm)		6.28	2.05		5.34	1.60	++
Glycogen µMol/g		12.08	12.06		29.37	13.40	+++
Lactate µMol/g		91.33	32.44		56.67	32.76	+++
ATP µMol/g		1.50	1.49		3.74	2.06	+++
R-value		1.21	0.22		0.93	0.24	+++

A further criterion for the difference between treatments
is the degree of muscling of the animal. When one group of pigs
is transported from one finishing stall it can be assumed that
the conditions of transport are comparable for all animals.
However, comparing the results of pH measurements from this one
group we see that the transport conditions affect meat much less
than conformation (Figure 3). The variation due to transport
with extreme difference in conditions was clearly less than
that due to carcase grade within treatments.

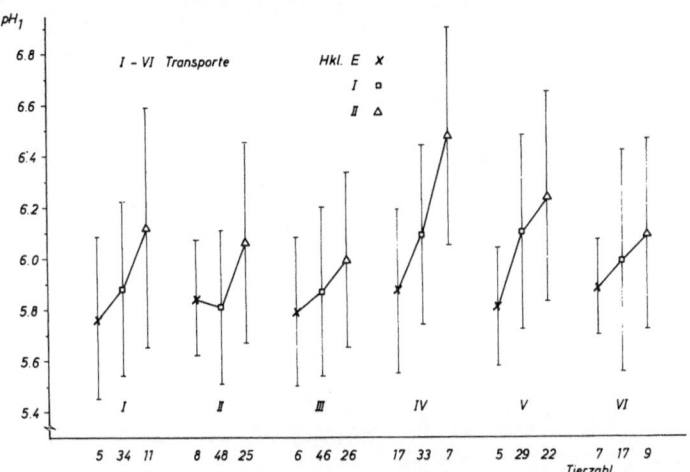

Fig. 3. pH value (*M. semimembranosus*) - transport - carcase grade.

DISCUSSION

E. Kallweit *(Federal Republic of Germany)* When you were talking about body temperature on arrival, was this before the animals were unloaded or afterwards?

Chr. Augustini *(Federal Republic of Germany)* It was immediately on the arrival of the truck and in the truck.

G. von Mickwitz *(Federal Republic of Germany)* We found the same as you did, for a small group. I think that under the practical conditions of a slaughter house you would find a good correlation between weather conditions and loading density on the one hand and body temperature on the other. It relates to the mass of 'meat' which is being transported. If you transport only 5 x 80 kg pigs there will be a difference in the heat production compared to when you transport a full load of slaughter pigs.

SESSION IV

MEANS OF TRANSPORT,
WITH PARTICULAR REFERENCE TO THEIR CONSTRUCTION

Chairman: J.P. Signoret

TRANSPORT OF DAY OLD CHICKS BY AIR

A. Hoogerbrugge and H.J. Ormel

Zootechnical Institute,
Veterinary Faculty, Yalelaan 17, Utrecht, Netherlands.

ABSTRACT

Sometimes a high rate of mortality occurs during transport of day old chicks. We know from practice that pre- and post-flight arrangements are very important factors but circumstances during flight may also cause death. We investigated some factors which might be of importance during flight. The main aim of our investigation was to answer the question: What is happening inside a chick box? The experiments were carried out in an aircraft and in a climatic cell. The most important results were:

1. chicks are very resistant to cold;

2. chicks can die within 45 to 50 min by high temperature;

3. high temperatures can only be prevented by good ventilation;

4. good ventilation or transport of warm air is only possible with:

   - a good design of chick box;

   - enough ventilation holes in the chick box;

   - enough space between the chick boxes;

   - enough space between the chick boxes and the bottom;

5. transport of warm air depends on the ventilation rate and ventilation system of the aircraft;

6. the ventilation system in most aircraft is not suitable for good ventilation during transport of chicks.

In conclusion, more research is necessary to prevent mortality during flight.

INTRODUCTION

Transport of animals by aircraft has increased considerably since 1945. At first transport was mostly concentrated on pet animals and zoo animals. Transport by air has many advantages when compared with transport by car, train or boat. It is not surprising that all kinds of industry have become interested in this method of transport. Nowadays horses, cattle, sheep, pigs, poultry and fish are transported by air all over the world.

Lack of experience and knowledge were the main reasons for many accidents which occurred, mostly characterised by a very high rate of mortality, especially in poultry.

TRANSPORT PROCEDURE

Generally 3 to 12 h after hatching the chicks are transported by truck to the airfield. On the airfield the boxes are put on pallets or directly loaded into the belly of the aircraft.

The time between loading and take-off has to be as short as possible for during loading there is mostly only natural ventilation in the aircraft. Some aircraft are equipped with extra artificial ventilation by an Auxiliary Power Unit (APU). During taxiing and during take-off there is only limited ventilation.

POSSIBLE CAUSES OF DEATH

There is a lot of speculation about the main cause of death. Some people believe that low temperature outside the aircraft during flight (-50° to -55°C), is responsible for high mortality. Others believe, especially in the case of charter flights, high temperatures or lack of oxygen to be the most important factors.

Low temperatures

In modern aircraft heating is present in all compartments. This means that only in the case of human or technical failure will the temperatures inside the belly drop to very low values. In addition Kaltofen and Dijk (1975) found in their experiments that day old chicks, kept single in boxes, did not show homoiotherme at that age. After a few hours in a cold environment body temperature had decreased to a value of about 20°C and after bringing the chick back into an environmental temperature of 35°C, body temperature had reached the normal level within an hour. Chicks kept together in boxes are able to keep their body temperature much longer by huddling.

The conclusion is that only in the case of mistakes or accidents is the death of day old chicks caused by low temperatures.

Lack of oxygen

During all our experiments both in the aircraft and in

the climatic cell we have found no indication which supports
the hypothesis that lack of oxygen might be a cause for death
of chicks during transport.

High temperature

Exporters and air transport companies, which have enough
experience, believe that in most cases high temperature is the
main reason for the high mortality. Important factors may be:
1. the large number of chicks which are transported on a
   pallet or a compartment;
2. the heat produced by the chicks;
3. the insulation rate of the chick box, the size and number
   of holes and the way the chick boxes are loaded (a very
   high density).
During flight we and Kaltofen (1975) have collected data on
temperature values inside the aeroplane. The data show a large
difference in temperatures found in boxes located at different
places and sometimes we found temperatures exceeding 35°C.
These high temperatures are dangerous for the chicks.

INSIDE THE CHICK BOX

It is not possible to do all kinds of experiments during
flight. For this reason we started a range of studies with one
or two boxes, firstly in an incubator and later in a climatic
cell.

Aircraft conditions were simulated by variations in
temperature, humidity, circulation and ventilation.

The figures show that:
- when the outside temperature of the box is over 34°C and
  ventilation is limited the temperature inside the box will
  increase rapidly (within 10 - 15 min) to values exceeding
  40°C;
- there is a strong correlation between body temperature and
  the higher inside temperature;
- when body temperature rises above 40°C the activity of the
  chickens rises, which means an increased heat production;
- at the high temperature of 40°C the chicks start panting;

- when ventilation is decreased the relative humidity
  inside the box very quickly reaches 90% or more which
  means that the possibilities of heat loss by evaporation
  are decreased;
- when body temperature reaches values of about 43° to 44°C
  (within 40 - 50 min) chicks become comatose and will die
  very quickly.

OUTSIDE THE CHICK BOX

The heat produced in the chick box can only be removed by
convection or radiation. As most chick boxes are surrounded by
other boxes the net heat loss by radiation is about zero. The
only possible way, therefore, to get rid of the heat is by
convection.

The loading configuration of the boxes depends on the
sizes of the boxes and on the compartment in the aircraft. To
prevent boxes from moving they are often stacked like bricks.
This means that air circulation between the boxes is limited,
and heat removal is reduced.

Much better ventilation is possible when there is an open
space between each stack of boxes, but this will give poor
adherence between the individual boxes, making the use of so-
called 'spacers' necessary to keep the box structure rigid.
How far the ventilation rate is decreased by the spaces depends
on the design of these spacers. Besides the space between the
boxes the movement of the warm air also depends on the space
between the lower layer of the boxes and the loading floor.
Optimum air circulation is only possible when the warm air in
the spaces between the stacks is able to rise. But this warm
air can only rise when there is a source of fresh air to replace
it. For a maximum replacement of warm air it is necessary that
fresh air enters from the loading floor, independent of the
chicks being loaded on the main deck or in the belly.

The transport of warm air depends on:
- ventilation rate of the aircraft;
- ventilation system;
- circulation between the boxes.

The ventilation rate of each type of aircraft is well known, though one has to take into account the fact that the capacity is not constant. Before and during take-off and before and during landing the ventilation rate is remarkably decreased. This means that there is a rise of temperature inside the aircraft during these periods. These periods of lesser ventilation have to be kept as short as possible, or a critical situation will arise.

Ventilation system

In most aircraft, perhaps in all, the fresh air inlet is situated at the top of the compartment and the outlets at the side wall or at the loading floor so the ventilation system is not suitable for transportation of large numbers of chicks. One can even say that the present ventilation system is not suitable for any animal transportation which involves a layered system of stacking. For chicks and poultry the removal of warm air is lower than the ventilation rate of the aircraft would suggest.

Ventilation rate

The following factors are important in improving the ventilation rate and to keep it constant under the given circumstances:
- enough ventilation holes in the chick box;
- a vertical space to be present for at least two sides of the box;
- the vertical space to be at least 7.5 - 10 cm wide depending on the number of boxes;
- at least 2 - 4 cm horizontal space between the boxes, depending on the design of the box;
- a space of 5 - 10 cm between the bottom of the lower box and the loading floor;
- no cross-obstacles in all spaces.

Some other factors

It is very difficult for the aircraft company to do the

job without mistakes. One of the most difficult factors is the fact that different types of chick boxes are used. During loading it is very difficult for the people involved to keep enough space between the boxes on the one hand and sufficient 'strength' of the stack on the other. When the stacking structure of the boxes proves to be too weak then the spaces between the boxes may be blocked at different places during the flight. The success of the transport, therefore, depends very much on the skill of the people. This means that most accidents occur in companies which have little experience and a lack of trained people. For the better companies are often very disappointed when they see what is happening on some of the airfields and during transport by truck in the country of destination.

Research

The present system of transport of day old chicks is more a result of practical experience than of research. Very little research is done in this field. We know also, partly from practice, that many factors are involved. Research projects may be:

1. the design of the chick box;
2. the spacers which can be used;
3. improvement of the ventilation system in the aircraft;
4. influence of light on the activity of chicks;
5. influence of age on the activity of chicks;
6. the influence of the weight of the chicks on the number of chicks which may be put in a box (under different circumstances);
7. the influence of drinking water or injection before transport.

REFERENCE

Kaltofen, R.S. and Dijk, D.J. Weerstand van eendagskuikens tegen verschillende combinaties van omgevingstemperatuur en luchtvochtigheid (Internal report, IPS 't Spelderholt).

DISCUSSION

W. Müller *(Federal Republic of Germany)* You told us that the limiting factor was not the pay load but the ventilation rate. Do you have the ventilation
rates per 1 kg of day-old chicks per hour and how many m^3 are needed?

A. Hoogerbrugge *(The Netherlands)* We do use these measurements, but I do not
have the figures at the moment. We normally calculate by chick boxes.

W. Müller I have also prepared a short paper about ventilation rates, and I
agree with your experiments. We made our own experiments in an aerosol chamber
and we calculated the ventilation rate via the metabolic rate and the maintenance metabolic rate because it is very difficult to get the requirements. One
airline company will tell us that they need one ventilation rate, and another
will have other figures, so we tried to calculate the rate. We found that
there is a difference between the flight operation and the ground operation,
because during flight the outside air is, perhaps, -50°C and this means 100%
relative humidity and a very low absolute humidity. If you heat this incoming
air to no more than 8-10°C the heat produced by the aircraft will mix with it
and over a distance of only a few cm the temperature will rise further. In
the boxes we use 34°C or more.

In flight, therefore, we need a lower ventilation rate than on the ground.
Some airline companies use additional ventilation systems for ground operations.
This additional ventilation must have the same rate as that on the aircraft.
If this is possible, there will be no problem with intermediate landing, except
in an airport where the ambient temperature is as high as 42°C.

My second question is whether you have any experience of the use of the
non-ventilated belly compartment for the transportation of day-old chicks?
Can you give us any figures as to the number of chicks, the time, the volume
and so on?

A. Hoogerbrugge I have both good and bad experience in relation to your last
question. One of our Dutch airlines fitted special ventilation in the belly
of aircraft for the transportation of animals, especially chicks, although
it can also be used for dogs, cats and zoo animals. Other airlines are not
aware of the need. As a result we encounter a lot of problems because of the
lack of ventilation.

It is possible to carry chicks in the belly without using artificial
ventilation, but you have to calculate the cubic capacity and the distance
of the journey before you know how many chicks you can carry.

I do not fully agree with your first point. For 30 - 45 min before landing the ventilation system decreases. The temperature increases and there is
very low humidity. There is no problem with long distance journeys, but there
are problems during landing and on landing if the airfield personnel are un-
aware of the requirements. Regulations are required to provide for the best
possible system of ventilation and prevent high temperatures being reached.

M. Jespersen *(Denmark)* It seems to me that the circulation inside the cabin
is created only by natural temperature differences. Has anyone tried to make
a more even distribution of the fresh air and a more forced circulation by
using piping?

A. Hoogerbrugge There is a forced circulation. Only up until now the inlet
has been on the top of the aircraft and the outlet on the bottom or the side.
That is wrong. The natural circulation goes one way and the forced circulation goes completely the other way. That is a conflicting system, and you also
see it in the stacks. When there is no real problem, you find that the highest
mortality is in the 5th and 6th layers, which is where the conflicting situation is in the stacks.

M. Jespersen Then there is no piping between the inlet and the outlet? It is flowing by itself.

A. Hoogerbrugge If you have piping, you have to have gaps in the pipes. We know from stables and houses that warm air will go upwards. We have not discussed this with designers, but they say that our system is good when we transport horses, cattle and people, and it is only when we transport chicks that we need another system. We are now trying for a system which could be used both ways, so that when you transport horses and cattle it is used one way and it is used the other way for transporting birds or chicks.

W. Müller I have a suggestion for calculating the weight of animals which can be transported in the belly compartment. We need only the CO_2 production of 1 kg of the animal and we can then obtain the total kg bodyweight which can be transported by a calculation involving the total volume of the belly compartment and the time of transportation. What is your opinion?

A. Hoogerbrugge I think that you can prevent problems with this system. For instance, in our situation, we had chicks in the belly hold. We made a theoretical calculation of the ventilation, but we had problems in one of the pallets, because there was a compressor near the pallet, producing a lot of heat. In other aircraft you might have other machinery which will produce heat. You can make the theoretical calculations but you have to be very careful to know exactly what is going on in the belly hold so that you might have temperature registration for the different parts of the belly. One part might be 30ºC while another part is 22ºC, and this would lower ventilation, so that you could say that you have a ventilation rate of 4 000 m3/h, but it might be much lower, depending on the way the chick boxes are loaded and the temperature difference between the chick boxes and so on.

W. Müller You are talking about a ventilated compartment. I asked you about unventilated belly compartments.

A. Hoogerbrugge I would advise that animals only be transported in ventilated compartments.

W. Müller You are right, but they use the Boeing 737 for all small distances in Europe and they have a good success rate.

A. Hoogerbrugge Yes, but then you can carry only a limited number of chicks.

BULK TRANSPORTATION OF FARM ANIMALS BY AIR AND VEHICULAR FERRIES

M.E.T. Watts

Divisional Veterinary Officer,
Ministry of Agriculture, Fisheries and Food,
Tolworth, Surrey, UK.

ABSTRACT

Current UK policy is based on scientific research and subjective professional opinion. Farm animals being transported in aircraft and in vehicles on ferries are covered. Space, environment and access are considered. Simple guidelines enable standards to be assessed rapidly under field conditions.

INTRODUCTION

Over the past 10 years there has been a considerable number of farm animals transported by air and in vehicles utilising the roll-on/roll-off (RO/RO) ferries to the continent from the UK. Reports have been made to the Head Office of the Ministry of Agriculture, Fisheries and Food by field veterinary staff and other veterinarians. Some exporters are very experienced and their valuable and reliable reports have been recorded. Veterinary officers also have submitted reports on the loading of over 100 consignments and accompanied at least 110 flights with cattle, calves, sheep, pigs and horses. A critical analysis of these reports and information has been condensed into quick reference documents.

Of necessity, the statistics and tables have been simplified for clarity and ease of use. Constructive guidance was required in many areas, but emphasis has, so far, been placed on a) adequate space, b) suitable environment and c) suitable access for stockmen to attend the animals.

Trade patterns and fuel costs have intensified the demand for moving groups of stock at the lowest unit cost possible. High stocking rates and less regard to the well-being of the individual animal are obvious commercial demands.

The comfort or well-being of each animal must be adequate at all stages of the journey - awaiting loading, loading, awaiting start of actual journey, all stages of journey and unloading. Stress, distress and injury can occur long before irreversible physiological changes, culminating ultimately in death, occur. Survival at the end of a journey is not exclusive evidence of adequate welfare. Stress occurs whenever there is a departure from the normal status quo. It is physiologically normal, but results in a complex chemical reaction giving rise to a metabolic rate above the basal level. When the stress makes unnatural or excessive demands, the physiological balance is upset and may manifest itself as physical or mental distress.

BASIC REQUIREMENTS OF ANIMALS IN TRANSIT

Space to stand and lie naturally

There is very little recorded data on the sizes of animals. Considerable variation occurs in the same species or age group. Only broad classification of size can be made in advance. Over-all height, length and width can be visually assessed during loading to ensure a natural posture. Some extra space must be provided to allow movement to adjust posture and the animals' relationship to each other (Lusk, 1968).

When animals lie down, the overall width and length usually increases. The process of lying or rising involves movements into an open space either beside, behind or in front of the animal. This makes another demand on the additional space needed by each animal. The natural, preferred stance during travel is also significant. Horses will stand for long periods, adult cattle stand for about 4 h and then 15% will become recumbent. After another 2 or 3 h, different animals will be recumbent. In general, calves and sheep will normally stand for 2 - 3 h and most will be lying after 6 h. Pigs prefer to lie almost immediately they have adjusted to the new surroundings.

The stocking density must relate to the duration of the journey and the normal behaviour of the species at different ages. This paper considers groups of 6 or more - a single animal would require at least double the area.

Pen size/group size

Ideally, individual pens are safest, but are very wasteful on space. Large groups of animals economically sharing spare space make the best use of floor area. There has to be a compromise. The more animals in a group, the greater the chance of physical damage from trampling or crushing during any violent movement. During acceleration or deceleration, animals in large pens have a greater distance to travel before being given rigid support by the pen wall. The energies of inertia are also greater and will aggravate crushing. A group size of between 10 heavy animals (over 250 kg) and 15 - 25 light (50 kg) animals is proposed, with a 5% tolerance. These would be accommodated in pens approximately 8 - 10 m^2 for adults and 6 - 8 m^2 for small animals.

Environment for animals in transit

The thermo-regulatory balance, the supply of oxygen and the removal of waste gaseous products must be maintained at all times, otherwise distress and ultimately uncontrollable physiological changes of a catastrophic nature can occur. The metabolic rate of an animal is adjusted to accommodate some variation in ambient temperature, food and water supply, perspiration and other heat loss processes. Outside a certain range, the body cannot adjust and there will be a physiological emergency.

Taylor (1978) indicates that at 10 000 ft the atmospheric temperature and pressure is below levels at which the body can adjust and alveolar gaseous interchange is inadequate. Above 8 000 ft the physiological balance is dependent on the aircraft's air conditioning.

The forward progression of road vehicles causes a forced ventilation or airflow. The ventilation apertures are designed to give adequate airflow. When the vehicle is stationary, there may be no air movement and the local environment around the animals may depart significantly from the 15°C norm of a thermoneutral environment. Taylor (1978) states this neutral environment, or comfort zone, has been found to be 4 - 25°C for

large animals and 10 - 23°C for small farm animals. According
to Pointer (1972), very young suckling animals have a limited
thermo-regulatory mechanism and 12 - 23°C is ideal. Under
normal conditions the body can adjust to a temperature 3 - 5°C
above and below these limits for short periods.

Humidity

The humidity of the ambient environment can seriously
affect the dynamics of latent heat dissipation. It can be aug-
mented by water and water vapour excreted as a product of
normal metabolism. Absolute humidity is an expression of the
actual weight of water in a given mass of air. There is a
maximum, or full saturated, weight of water which can be
suspended in air and is related to the temperature of the air
and can be tabulated:

TABLE 1

MAXIMUM WATER VAPOUR IN AIR (SATURATION) (MILLER, 1977; BRUCE, PERSONAL
COMMUNICATION)

°C	mg H_2O/l	lbs H_2O/lb air
0°	4.8	0.004
10°	9.4	0.008
20°	17.3	0.015
30°	30.3	0.027
40°	51.0	0.045

Within the comfort zone previously suggested, at the cool-
est temperature of 5°C the air is 100% saturated with 6.8 mg
H_2O/l, but at 25°C one litre of air can contain 23 mg H_2O. At
30°C this 23 mg H_2O would cause the air to be 77% saturated.
There must, therefore, be a compromise between high temperatures
to accommodate a larger mass of water vapour and an environmenta
temperature where the animal body is in homeothermic equilibrium
resulting in the minimum production of water vapour from metabolism or latent heat dissipation.

The ambient humidity varies considerably throughout the
world. During the winter months in the UK it is often

90 - 100%. In other countries the norm is 0 - 20%. This has
a considerable effect on the environmental conditions which
develop during loading and take-off of aircraft and in the
stationary air on the vehicle deck of a roll-on/roll-off ferry.
Stevens (1980) refers to a Livestock Weather Safety Index (1968)
which shows the interrelation of temperature and humidity
(Figure 1).

Muller (1978) quotes DIN 18910 giving minimum ventilation
rates which would result in no build-up of the main metabolic
by-products, water vapour, CO_2, NH_3 and H_2S.

TABLE 2

	$m^3/kg/h$
Horses	0.16
Adult cattle	0.17
Calves	0.24
Sows	0.19
Store pigs	0.20
Sheep	0.23
Chicks	1.38

Bruce (personal communication) also quotes a minimum
ventilation rate of 0.6 m^3/sec/18 000 kg of cattle contained
in a pen 3.7 x 2.44 m where the ventilation is by convection
only.

Access

The welfare of animals during transportation is dependent
on preserving, throughout the journey, safe penning, 'comfort
zone' environment, a natural stance and no excessive aggression.
Thus animals need to be observed regularly by the stockman.
Feeding and watering, first aid and other treatment including,
on rare occasions, emergency slaughter will require the attendant to gain access to each animal.

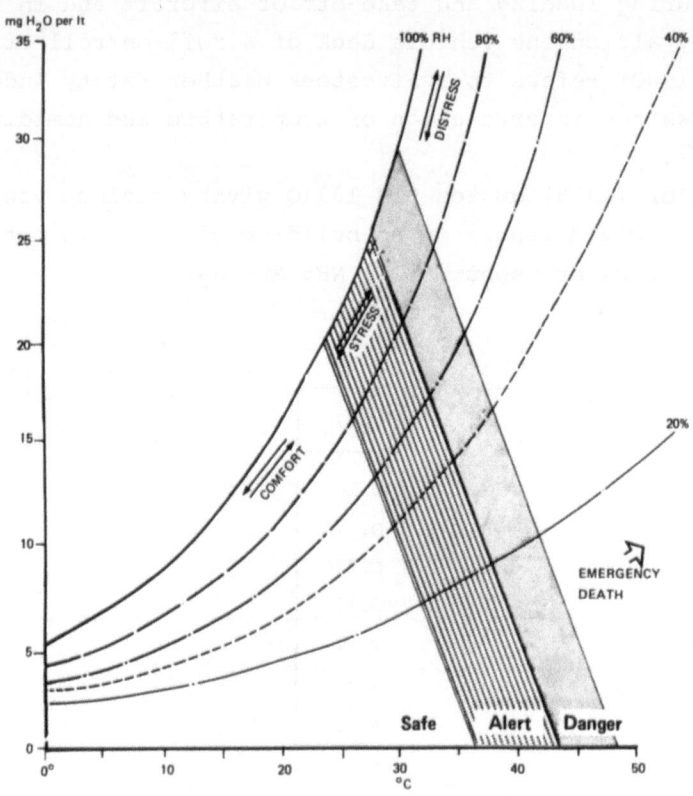

Fig. 1. Livestock weather safety index (after Livestock Conservation
        Inc Chicago).

Visual inspection requires adequate permanent or temporary
lighting. The attendant, if he is to make frequent inspections,
needs an easy, unobstructed and convenient means of gaining
access to each pen to view the animals contained in it. Walking
alongside of pens is the easiest. With two-tier vehicles, a
secure ladder to the upper tier pens is required. This should
have a step below the bed of the lorry to avoid a very high
first step. Secure hand-holds and footings are required on this
ladder and lateral steps from it to reach other pens or doorways.

On aircraft, a side walk passage utilises valuable floor
area. A crouch or crawlway over pens is sometimes provided.

Crawling is an unnatural stance which makes excessive demands on the attendant's knees etc. Since aircraft cabins can be 150 m long, this method of access is distressing to even a young, agile person. Crouch walking is slightly better but there must be adequate height, space and the minimum of physical obstructions. A trolley system running on a track over pens or suspended from roof rails gives easy visual access to all pens. The ingress and egress from the suspended, metal floored trolley requires some agility. Provided it is propelled gently over the pens, animals are not alarmed. There are air safety requirements for such a trolley and any access method must allow the attendant to have either emergency oxygen and contact with the Captain, or allow the attendant to return to his seat rapidly if air conditions are adverse in the opinion of the Captain.

To attend individual animals, easily operated access doorways into each pen are required. These must be safe and secure for both animals and attendant. During use animals should not be able to escape and the stockman's method of entry must be personally safe avoiding a physically difficult entry and any dangers from the stock within the pen.

CARRIAGE OF ANIMALS BY AIR

Load stability and trim

These are of paramount importance to the air personnel. For take-off, periods of turbulence and landing, the closer the animals are together, the better the mutual support. Large pens allow spare space to be shared with more animals. During periods of violent movement, however, there are more animal bodies which can squash together so leaving more empty space. This could upset trim and may cause excessive compression of animals underneath or at the front of the line of movement. Small pens reduce these risks.

Trim will often demand uneven load distribution. The stocking density should be geared to the heaviest payload areas and the demands of trim may result in lower densities and spare space in certain pens.

Pen construction in aircraft

The aircraft frame must be protected from physical damage by animals and from the corrosive action of excreta. Physically strong pen wall construction is often bulky and can cause interference with airflow patterns to such an extent that ventilation is impaired. Sheeting the outside of the pen or the lower areas of the wall of the aircraft with metal or plastic is the normal way of deflecting urine and faeces into the pen and away from the aircraft structure. Cabin input ventilators, if at waist height and, more frequently, exhaust ports near the floor, are thus obstructed. There is often a flow of air along the cargo cabin. Any high solid pen partitions obstruct this flow.

The simplest penning is constructed of aluminium piping 7 cm in diameter. The location of two, three or four horizontal rails, pen sizes and structural strength is very variable. Cargo aircraft cabins are between 2.25 m and 3.25 m internal width (except the Jumbo DC10 and 747 which are approximately 5.6 m wide). These 'tubular' pens span the full width of the fuselage and are between 2.5 m and 5.0 m long. Structural weakness is due to length of unsupported piping and stresses caused by movements of the animal cargo in the pens. The gaps between the horizontal pipes are often filled with weldmesh sheeting. This can be insecurely fixed to the rails or be fabricated of thin wire which cannot withstand the weight of animals thrown against it. This penning gives the minimum of protection to the aircraft from faeces and urine. Plastic sheeting usually forms an impervious floor and lower wall. This may be slippery, easily torn and also cause obstruction to the ventilation system.

There is no upward restraint and crew air safety is dependent on the forward 9 G net. The British Civil Aviation Authority now requires plans of such a construction and an aircraft to be fitted out before it is approved by them for air safety reasons. Each airline and model of aircraft will thus have a consistent design.

Fixed penning involves a temporary third skin to the aircraft with transverse full width pen gates attached to this

structure. It makes the most economic use of the space in
the cargo cabin, but cannot be easily dismantled and requires
specialist rebuilding personnel. Unless dismantled, it creates
areas difficult to clean and disinfect thoroughly.

Frequently, the penning system is based on the 2.24 m x
2.74 m or 3.18 m (88" x 108" or 125") cargo pallet. This,
being standard aircraft equipment, is suitable for the normal
handling equipment such as roller or moving tracks, dollies,
scissor lifts and fork lifts, available on cargo aircraft and
at airfields. Structural strength is restricted to one pallet
unit. Each pen is relatively small. The pen can be dismantled,
easily cleaned and disinfected and, if necessary, stored or
transported in a collapsed form.

Many palletised metal penning systems are designed by
engineers with a background of air frame strength and safety in
mind. These are strong and rigid, but may not make due allow-
ance for air flow, ease of loading animals, ease of access to
animals at all times and adaptability for species and size.
Penning systems not designed and built by aero engineers often
lack rigidity, being insecure for the animal and likely to
collapse as a result of movements of the animals or during
turbulence. Some airlines and exporters are very competent at
constructing palletised wooden crates which are disposable after
being used once. Much more care has to be taken with rigidity,
security of animal in pen and avoiding projections likely to
injure an animal.

Two tier penning suitable for the smaller farm animals
involves a solid floor forming a ceiling over the lower deck
at a height of about 100 cm. There is thus very little free
air space above sheep and pigs, and only small calves (50 -
60 kg bodyweight) can be carried in this limited headroom.
(Ashly, 1979). Ventilation of this almost enclosed lower deck
is critical and may necessitate reduction of the stocking density. Access, throughout the journey, to individual animals
in this lower deck may be difficult or impossible.

Space available for animal payload

Aircraft data sheets list the maximum floor area and

height available for cargo units. Similarly, the maximum overall size of any cargo unit is required by airline staff to
arrange its loading and subsequent location in the cabin.
These two quoted dimensions are not a true indication of the
space available for the animal cargo. Gross cabin floor area
is reduced by the gallery, Safety Spider Web, extra passenger
seats, taper or contour of tail, door ramps and control gear,
access passageway, thickness of pen wall or gate construction,
wasted aircraft space at the edge of a pallet. Animals cannot
utilise the full area of an 88" x 108" or 125" pallet. The
sloping contour of the fuselage dictates that when a two tier
system is possible, because of species, size and aircraft
ventilation capability, the top tier will have less utilisable
floor area - animals cannot stand in the angle of the contoured
side wall.

A total pallet area of 7.1 m^2 occupies a floor area in
the aircraft of 7.75 m^2. The construction of a metal or wooden
pen within the fixing (Vickers) track on the pallet reduces
the available floor area for stock to 6.3 m^2. In a fixed penning system, side walls and gates of pens are 5 - 8 cm thick
and nothing fits the fuselage exactly so reducing gross floor
area to a small usable area.

Aircraft ventilation

Allsup in 1977 gives a comprehensive summary of aircraft
currently being used for animal transportation. They can be
roughly classified into 3 groups:
a.   short haul unpressurised propeller models;
b.   medium/long haul pressurised turbo propeller models;
c.   long haul jet engined models.
The capacity and sophisticated control of the ventilation
system vary from a through draft of ambient air in (a) to 3
compressors, of considerable potential, linked with heat inter-
changers and a freon cooling system in (c). A water and CO_2
absorption system may also be included in the air recirculation
system.

When an aircraft is stationary, its cabin environment is
dependent on an auxiliary ventilation system. This can be an

internal power unit, an external power supply activating the
aeroplane's own air conditioning unit or supplementary exhaust
fans or an external air conditioning unit. Neither of these
systems has the capacity or capability of the full air conditioning powered by at least one and maybe three aeroengines. During
initial ascent and final descent, 80% of the aircraft's ventilation capacity is utilised, for air safety reasons, by the
engines and hence only a fifth is available for the cargo cabin
for periods of 10 minutes or so. This has been demonstrated
by Muller (1978) and others. Any recirculation of air without
absorption will cause a build-up of CO_2 and water vapour.
Flight engineers usually combat the problem of a dewpoint by
raising temperature.

The distribution of hot and cold air may not be even
throughout a cabin and also may not be standard for a model of
aircraft. A further complication is that the electronic sensors
are situated in the ventilation trunking near the skin of the
aircraft and may not give the engineer a true value for the
environment in the pens near the animals.

Airlines and manufacturers quote the air conditioning
capabilities for an aircraft in level flight, fitted out as a
passenger aeroplane. Cargo aeroplanes always have some mod-
ifications to the interior and this can sometimes drastically
reduce or produce an uneven distribution of ventilation air.
Specifications do not cover stationary aircraft, temperature/
dewpoint problems caused by urine and heat output by nervous
animals. In the UK the ambient humidity and poor ventilation
of the cargo cabin during loading almost invariably means the
cabin humidity is 100% RH at take-off. The use of supplementary
ventilation in the form of ground power unit and/or ground air
conditioning unit and/or rear passenger doorway fans is thus
almost a standard requirement, otherwise cabin conditions at
take-off are beyond levels which can be rectified by the aircraft's own air conditioning pack during subsequent level flight.

Lower belly holds also take palletised/containerised
cargo. These smaller holds may or may not be pressurised.
Pressurisation is from a spill-over of secondhand air from the
passenger cabin or cargo cabin above. There is very limited

control over its temperature and gaseous content. Exhaust is
usually via a small leak hole in the door. The standard
specification of a change of air in a passenger cabin of at
least every 3 minutes is reduced to two or three changes per
hour in these pressurised holds. The environment of the lower
hold is further influenced by heat or fumes generated by aircraft equipment (particularly batteries) associated with the
hold and the inert volume and cryogenic effect of other cargo
in the hold. 'A' classified holds may be suitable for a small
number of animals, but total oxygen requirement, overall
humidity and temperature, must be assessed for the complete
journey. Stocking density cannot be related to floor area for
belly hold cargoes.

Stocking density guidelines for aircraft

An assessment has been made of the opinion of many
veterinarians on numerous flights. The graphs in Figures 2,
3, 4 and 5 have been edited for clarity.

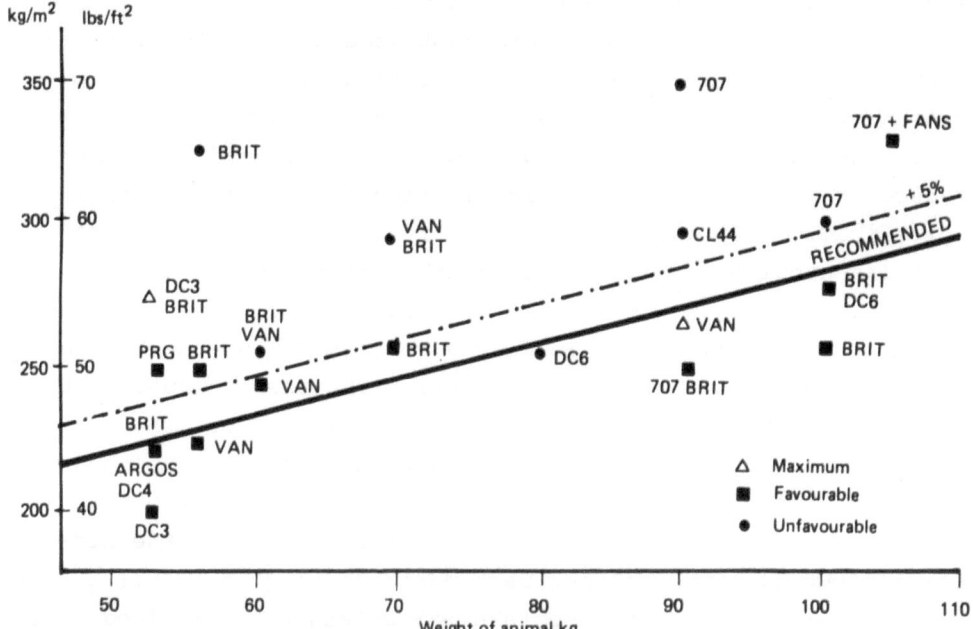

Fig. 2. Observed stocking densities in calves.

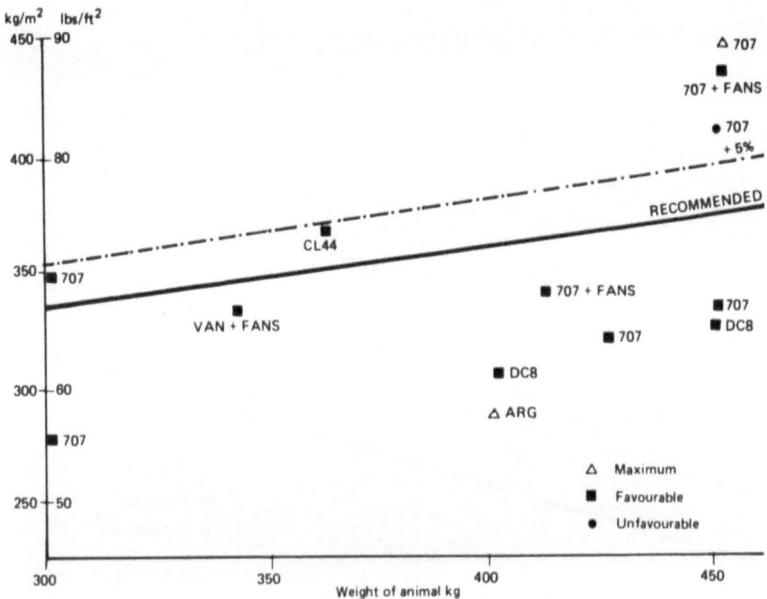

Fig. 3. Observed stocking densities in adult cattle.

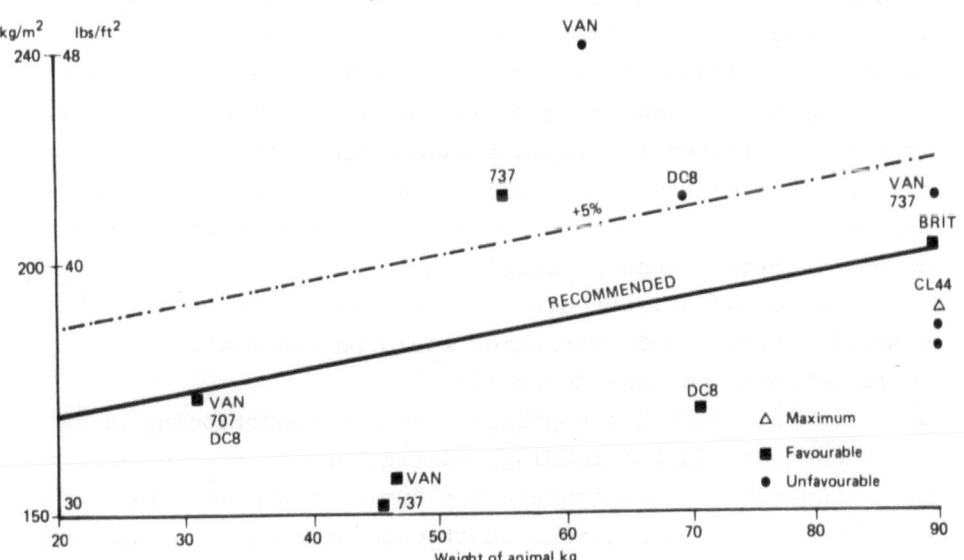

Fig. 4. Observed stocking densities in pigs.

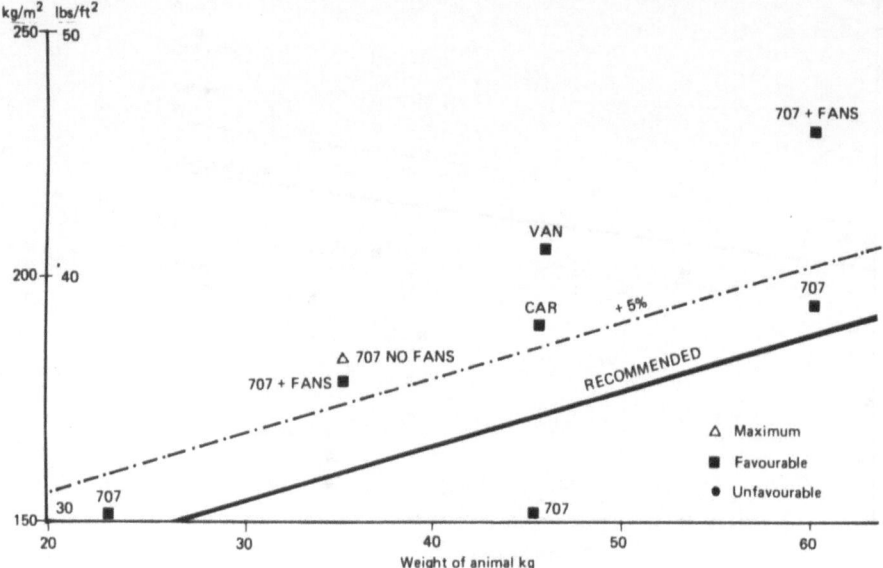

Fig. 5.   Observed stocking densities in sheep.

Benedict's (1938) correlation of products of metabolism
with surface area is impractical under field conditions.
Muller (1979), quite rightly, relates ventilation requirements
to a complicated mathematical interrelation between cabin
humidity and temperature, ambient humidity and temperature and
water vapour production by a full load of animals. This again
is too complicated for field application. There is a log-
arithmic correlation between body weight/surface area and
production of metabolic by-products. Non-scientific personnel
cannot interpret such a correlation.

The UK guidelines are a considerable compromise and
simplification. A 5% variation would be reasonable. An
increase could be considered if:

a.   the aircraft has sophisticated air conditioning which is
      adequate during loading, waiting and unloading, or

b.   the ambient temperature and humidity are low, or

c.   no access is required to attend animals, or a crawlway or
      trolley system is provided, or

d.   there are no legal constraints/stipulations on space or
      other welfare matters, or

e.   the total journey time is short - less than 4 h.

The stocking density should be reduced if:

a.   there is inadequate ventilation during loading, waiting or unloading, perhaps due to poor ground support equipment, or

b.   the upper limits of the comfort zone ($25^{\circ}C$ 80% RH) will be reached before the aircraft's system becomes operational, or

c.   unimpeded access to pens and reasonable access to stock is not possible, or

d.   special segregation of groups or species for reasons of variation in age, density of fleece, stage of pregnancy, compatibility, or

e.   journey time over 4 h and 100% lying space required, or

f.   the volume, thermodynamics or other hazard pertaining to other cargo in cabin or belly hold demands review and assessment.

TABLE 3

STOCKING DENSITY GUIDELINES FOR AIRCRAFT

Species	Weight kg (other weights pro-rata)	lbs/ft^2	kg/m^2	Space/animal ft^2	m^2
Calves	50	45	220	2.45	0.23
	90	55	270	3.6	0.34
Cattle	300	70	344	9.0	0.84
	500	80	393	13.75	1.27
Sheep	25	30	147	1.8	0.17
	70	40	196	3.85	0.36
Pigs	25	35	172	1.6	0.15
	100	40	196	5.5	0.51
Horses	400	60	295	15.0	1.36
	500			18.0	1.7

These densities can be expressed as a graph (Figure 6).

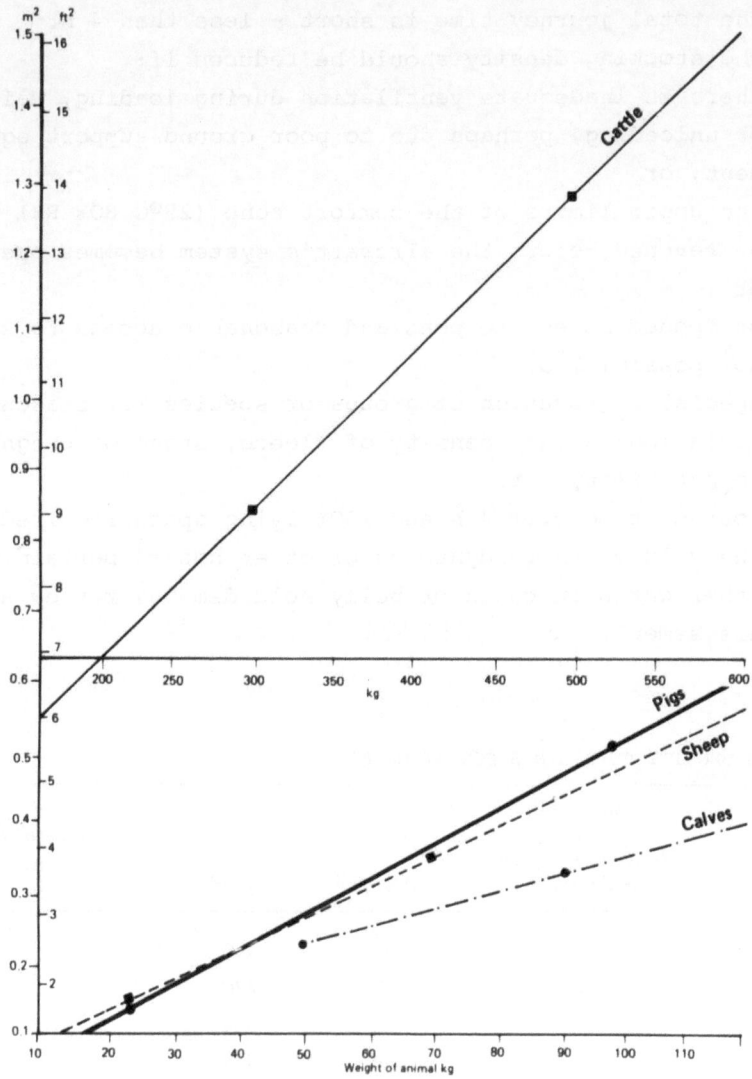

Fig. 6.   Optimum aircraft stocking density floor space/body weight.

CARRIAGE OF ANIMALS BY VEHICLE ON FERRIES

Ventilation of lorries on RO/RO ferries

The UK State Veterinary Service has considerable experience
in the transport of the smaller farm animals in specialist two-
tier vehicles utilising the ferries crossing the North Sea or
Channel to the continent of Europe.

Although ferry decks have forced ventilation giving about
20 changes per hour, there is no appreciable air movement and no
forced ventilation in the vehicles. The air change suggested
above to maintain 'comfort zone' and safety is thus motivated
by convection only. Research and subjective observations
indicate that a 4 - 6°C temperature differential will create an
adequate airflow, provided there are relatively large ventilation
openings on each side of each pen on each tier of the lorry.
A single unobstructed ventilator near the roof at least 49 cm
deep for the full length of the pen is required. If there is
a low level (near floor) full length inlet ventilator, then
the same air change will occur with a smaller upper or 'ceiling'
ventilator (Bruce, personal communication). This has been
confirmed by the monitoring by Veterinary Officers of air flow,
temperature, CO_2 and NH_3 levels throughout ferry journeys of
4 - 6 h duration. Full length lower ventilators greater than
10 cm in depth do not appreciably increase the air flow and
their additional depth can be ignored.

These ventilation apertures perform the same, irrespective
of the overall height of each tier. The areas can also be expressed as a percentage of the total side area of the pen or lorry
side. These minimum requirements can be tabulated (Table 4).

Obstructions such as framework ribs greater than 10 cm
wide reduce the effectiveness of the ventilator and will necessitate an increase in depth of the individual openings. Similarly, the access door may provide additional ventilation to one
side of one pen, but there must be some provision to prevent an
animal falling out if the door is opened for ventilation.

These ventilation apertures are excessive for UK road use.
Some means of closing is also required to reduce the area to 8%,
approximately, when the lorry is used on a cold, damp day for

TABLE 4

MINIMUM VENTILATION REQUIREMENTS

Tier height (m)	Single top vent	Minimum depth of full length unobstructed vent cm											
		Bottom and top vents at least 30 cm apart											
		B	T	%	B	T	%	B	T	%	B	T	%
1.22 to 1.36	49 cm 40%	2.5 cm	38 cm	33	5 cm	29 cm	28	7.5 cm	23 cm	25	10 cm	19 cm	24
1.37 to 1.51	" 35%	"	"	29	"	"	24.5	"	"	22	"	"	21
1.52 to 1.9	" 30%	"	"	24.75	"	"	21	"	"	18.75	"	"	18

example. The low level ventilators should be closeable, because of the possibility of wind chill on small animals.

Stocking densities on RO/RO ferries

The vehicles are all subject to UK domestic legislation whilst travelling on roads to the docks. Road vehicles rarely exceed approximately 2.5 m (8'2") in width and internal pen lengths must not exceed:

for adult cattle	3.7 m
calves	2.5 m
sheep	3.1 m
pigs	3.1 m

Thus, in any one pen there is a floor area of about 9.25 m^2 for adult cattle and 6.25 m^2 for calves. UK Sea Transport legislation also requires pens to be 8 - 9.5 m^2. Experience has shown that for journeys less than 12 h the ideal densities are those shown in Table 5.

If the total journey by road and RO/RO ferry exceeds 12 h, the densities shown in Table 5 should be reduced by 25% to allow more space for lying and for feeding and watering.

TABLE 5

STOCKING DENSITY ON FITTED AND RO/RO VESSELS

	Density		Space/animal		Maximum number/ pen
	lbs/ft^2	kg/m^2	ft^2	m^2	
Cattle:					
175 kg	74	365	5.2	0.48	20
500 kg	78	384	14.0	1.3	7
Calves:					
50 kg	31.5	154	3.5	0.325	20
75 kg	33	163	5.0	0.46	13
Pigs:					
50 kg	40	192	2.75	0.26	30
75 kg	40	197	4.1	0.38	20
Sheep:					
50 kg	30	147	3.7	0.34	23
75 kg	30	147	5.5	0.51	15

REFERENCES

Allsup, T.N., 1977. Recent Developments in the Transport of Farm Animals
    and Horses. Veterinary Record 100; 211-216.
American Society of Heating, Refrigeration and Air Conditioning Engineers,
    1977. In: Handbook of Fundamentals. New York (Quoted by Miller).
Ashby, B.H. et al., 1979. In: Advances in Agricultural Technology. United
    States Dept. of Agriculture. (ref. AAT NE.5).
Bell and Davidson, 1952. Physiology and Biochemistry 2nd Edition.
    E. and S. Livingstone, Edinburgh and London.
Benedict, 1938. In: Study of Basal Metabolism No. 503. Carnegie Institute,
    Washington.
Bruce, J.M. Personal communication. Aberdeen.
Dragerwerk, A.G., 1975. Technical Manual 11th Edition. D 2400 Lubeck 1.
Livestock Conservation Inc. In: Livestock Weather Safety Index Conference.
    Chicago. (Quoted by Miller).
Lusk, G., 1968. Science of Nutrition. 1st Edition. Saunders Press,
    Philadelphia.
Miller, R.R. In: United States Department of Agriculture. Staff Manual.
Müller, W., 1978. In: Climatic conditions during air transport of farm
    animals. University of Horenlein, Stuttgart.
Pointer, J., 1972. In: Proceedings of the Institute of Agricultural
    Engineers Conference 1972. London.
Stevens, D.G., 1980. Ventilation for Air Transport - Transport Marketing
    Research Unit. College Station, Texas.
Taylor, G.B., 1978. Transport of Animals by Air. IAS Cargo Airlines
    Manual. Gatwick.

TRANSPORTATION OF SHEEP BY SHIP FROM AUSTRALIA
TO THE MIDDLE EAST

W. Müller

Institut für Tiermedizin und Tierhyqiene (460),
Universität Hohenheim, Postfach 700562, D-7000 Stuttgart 70,
Federal Republic of Germany.

INTRODUCTION

During the last few years transports of sheep livestock from Australia to the Middle East, especially to Saudi-Arabia, Iran and Kuweit, have become increasingly important.

Some shipping companies have specialised in sheep transports while using converted passenger ships but tankers are also used. Normally the number of animals varies between 30 000 to 80 000 animals per transport. Generally the transports start from West Australia.

The passage lasts 11 to 16 days depending on the speed of the ship. By reason of the present high oil prices the passage tends to be of a longer duration because it is more economical to sail at a slower speed. During all observed voyages the total loss rate lay between 1.5 and 8%.

In Table 1 the daily losses and losses estimated by means of a power function are listed for a journey which I accompanied myself. The Table shows that when r^2 = 0.94 a high degree for the quality of adaptation is reached.

$$y = a \cdot x^b$$

y = daily losses
x = day of journey
a = 5.2
b = 1.29

On this passage 116 dead animals were dissected between the 3rd and 16th day of journey, and the results of the dissection are shown in Table 2. One can notice that of the 116 animals 112 had alterations of the kidneys which are characteristic of pulpy kidney disease. Two animals showed alterations of the lungs and two more had pericarditis.

TABLE 1

LOSSES OF ANIMALS DURING THE VOYAGE

Day	Total	Daily	Calculated estimation $y = a \cdot x^b$
1	8	8	5.2
2	18	10	12.71
3	30	12	21.46
4	54	24	31.11
5	94	40	41.49
6	154	60	52.50
7	239	85	64.05
8	330	91	76.10
9	439	109	88.60
10	555	116	101.50
11	692	137	114.79
12	816	124	128.43
13	926	110	142.41
14	1 078	152	156.70
15	1 275	197	171.30
16	1 421	146	186.18

y = losses $a = 5.2$ $r^2 = 0.94$
x = day $b = 1.29$

ANIMALS WITH KIDNEY LESIONS

During the dissection all stages of kidney lesions were found. After removal of the kidney capsule a segmental swelling could be seen. The tissue was infiltrated with blood and often petechial haemorrhages.

Sometimes only parts of the kidneys were changed (showing a deep red colour and easily squeezed with a finger). If the organs were left in the cadaver one could observe an autolytic decay of the whole organ after a few hours.

Other animals showed a yellowish discolouring of the kidney (like being cooked) with some one third to one half of the organ covered with deep red spots which could be squeezed by pressure with the finger.

TABLE 2

SECTIONS FROM 116 SHEEP WHICH DIED DURING THE VOYAGE

Day	Number of animals	Pulpy kidney disease in x animals	Other diagnosis
2	6	6	-
3	8	7	Pericarditis
4	7	7	-
5	8	8	-
6	8	8	-
7	-	-	-
8	9	9	-
9	10	10	-
10	7	7	-
11	7	7	-
12	10	8	Pneumonia Pericarditis
13	9	9	-
14	12	12	-
15	15	14	Pneumonia

The liver generally had slightly swollen edges. The spleen was always sharply edged. Animals with an abnormally full urinary bladder also had an abnormally full gall bladder, and if the urinary bladder was not very full, the gall bladder was empty or only slightly full.

A total of 4 animals showed gas accumulation under the fascia of the *M. gracilis* beside the kidney changes.

The lungs were without special changes. Newly dead animals showed no fluid accumulation in the pericardium, animals which died the previous night had 1 - 2 thimbles of fluid in their pericardium.

In most cases the mucous membrane of the abomasum was diffusely reddish, as was the intestine. The intestine also showed gas formation and its vessels were deeply injected. These alterations were not only found when animals died but also when seriously ill animals were killed.

The symptoms occurred abruptly. The animals lay on one side and tried to get up while paddling with their limbs. The head of some animals was bent to the neck. There was no spasm of the muscles. Diarrhoea (sometimes bleeding, but mostly without blood) could only be noticed in some animals briefly before death.

DETECTION OF SALMONELLA

Intestine samples were collected from 48 dissected animals and incubated in Na-Tetrathionat-Broth for 16 h (37°C ± 1°C). After this a loop of the incubated material was inoculated on brilliant green-phenol red-lactose-agar and incubated for 18 h (37°C ± 1°C).

In all 48 samples only one germ was lactose-negative, but could not be agglutinated with polyvalent salmonella serum. The detection of salmonella was negative for all 48 samples.

Besides the salmonella test, intestine samples of 47 animals were incubated in standard-I-broth for 16 h and afterwards incubated on standard-I-agar for 18 h. The isolated germs were inoculated in 'stitch-agar' and were taken to the laboratory in Stuttgart. After purification of the germs their resistance against different antibiotics was tested.

VENTILATION RATE

During such journeys all temperature zones are crossed and this is why the ventilation system must be very adaptable (Table 3).

Table 4 shows the different demands on the ventilation system during such a journey (theoretical relevance).

SPACE REQUIREMENTS

The floor space which was provided for one animal varied between 0.3 m^2 and 0.33 m^2 and we considered this to be sufficient.

The strewing of superphosphate reduced the NH_3-concentration in the decks (Figure 1) and experiments during another transport showed that sheep losses caused by pulpy kidney disease can be reduced by the strewing of superphosphate

TABLE 3

CLIMATIC CONDITIONS DURING THE VOYAGE ESPERANCE - DJEDDA (SEPTEMBER 1979)

Day	Outside Temp. °C	RH %	E-deck Temp. °C	RH %	NH_3 (ppm)	CO_2 (Vol. %)
1	10	50	21	65	–	0.08
2	14	52	22	70	10	0.08
3	16	60	23	80	7	0.08
4	22	50	24	90	20	0.08
5	23	50	25	90	30	0.08
6	23	80	28	95	35	0.08
7	25	82	30	90	30	0.08
8	28	65	30	91	30	0.08
9	30	85	31	90	35	0.09
10	31	70	32	92	25	0.09
11	30	80	33	91	15	0.09
12	30	80	33	90	15	0.09
13	28	80	30	95	15	0.09
14	29	80	33	90	15	0.09
15	34	90	36	90	15	0.09
16	35	90	36	95	20	0.1

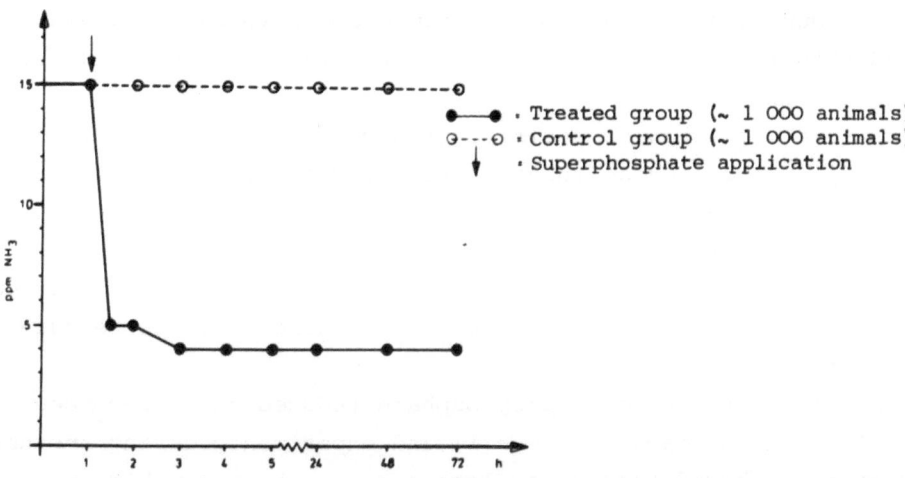

Fig. 1. NH_3 concentration after application of superphosphate at 33°C and 90% relative humidity (E deck, 100 kg/1 000 animals).

TABLE 4

AIR REQUIREMENTS FOR SHEEP (45 kg)

Temp. (°C)	Fresh air RH %	Minimum air flow (m^3/h kg)	Resulting CO_2-concentration (Vol. %)
A) Winter conditions		Deck: 18°C, 80% RH	
4	60(80)	0.1945 (0.2132)	0.2840 (0.2591)
6	60(80)	0.2185 (0.2424)	0.2528 (0.2278)
8	60(80)	0.2458 (0.2854)	0.2247 (0.1935)
10	60(80)	0.2854 (0.3403)	0.1935 (0.1623)
12	60(80)	0.3403 (0.4537)	0.1623 (0.1217)
14	60(80)	0.4213 (0.6554)	0.1311 (0.0843)
16	60(80)	0.5708 (1.1797)	0.0968 (0.0468)
18	60	1.1060	0.0499
B) Summer conditions		Deck: 24°C, 80% RH	
18	60(80)	0.2809 (0.3765)	0.1966 (0.1467)
20	60(80)	0.3277 (0.5530)	0.1685 (0.999)
22	60(80)	0.4316 (1.0409)	0.1280 (0.0531)
24	60	0.6554	0.0843
C) Tropic conditions *		Deck: 34°C, 90% RH	
28	60(80)	0.1578 (0.1945)	0.35 (0.2840)
30	60(80)	0.1623 (0.2528)	0.3403 (0.2185)
32	60(80)	0.1945 (0.3687)	0.2840 (0.1498)
34	60(80)	0.2424 (0.7078)	0.2278 (0.0780)

* For the elimination of NH_3 it is necessary to use 0.7078 m^3/h/kg.

(Kormoran, Table 5). One year later another experiment showed no influence of superphosphate, but also no influence of a vaccination programme on the number of losses which indicates that beside the losses caused by clostridia other circumstances must play an important role.

If *Cl. welchii* is involved in the disease of sheep, it must be concluded that the observations indicate an infection through the following mechanism: the 'infective agent' is

TABLE 5

MORTALITY OF SHEEP (VOYAGE 7/79 PORT ADELAIDE TO DJEDDA)

	Number of animals	Mortality
No special treatment	1 451	2.27%
Vaccinated against pulpy kidney disease	3 100	0.45%
Superphosphate 100 kg/1 000 animals	2 600	0.11%

brought onto the ship by some of the animals or by the feeding stuff and causes the disease in a few sheep.

The germ is eliminated, mostly through excretion which now represents a source of infection for the other animals. The concentration of the germ in the excrement increases. because there is no manure management (objectively it is not necessary). The offered feed (pellets) produces good conditions for a disfunction of the rumen and abomasum and the possibility of a passage of spores, for example, into the intestine with a resultant toxin production and starting of the disease. In order to clarify the proceedings during the infection invest-igations were carried out in the laboratory.

The spraying of vegetative germs in an aerosol chamber and the attempt at recultivation failed, as the germs did not survive the airborne state. Spores, however, could be isolated from the airborne state without difficulty and the biological half-life period 6.24 h (β biol = 3.085 10^{-5}) could be estimated.

Because only vegetative clostridia can produce toxins, the sick animals must eliminate vegetative forms, but the chance of survival is only raised if the clostridia are sporulated, so the 'infective agent' must be sporulated before it can represent a source of infection for other animals.

In an experiment the role of the superphosphate on the sporulation rate was investigated. Different concentrations of superphosphate were added to clostridia cultures in sheep man-ure. The results are shown in Table 6.

TABLE 6

ELIMINATION OF VEGETATIVE *CL. WELCHII* IN EXCREMENTS OF SHEEP

Days	1 Veg.	Spores	2 Veg.	Spores	3 Veg.	Spores	4 Veg.	Spores	5 Veg.	Spores	7 Veg.	Spores	8 Veg.	Spores	9 Veg.	Spores
Control n = 3	+	-	+	+	+	+	+	+	+	+	+	+	-	+	-	+
0.2% SP n = 2	+	-	+	-	+	-	+	+	+	+	+	+	-	-	+	-
1% SP n = 2	+	-	+	-	-	-	-	-	-	-	-	-	-	-	-	-
3% SP n = 2	+	-	+	-	-	-	-	-	-	-	-	-	-	-	-	-
6% SP n = 2	+	-	-	-	-	-	-	-	-	-	-	-	-	-	-	-

It can be noted that superphosphate in a concentration of 1% and higher can inhibit the sporulation of vegetative clostridia, and this leads to the quick elimination of the germs in the faeces. Within two days the germs are no longer detectable.

CONCLUSIONS

1.   Because of the high temperatures of the loading space and the high outside temperatures a ventilation rate of 0.7078 m^3/h/kg is necessary for sea transport in tropical regions.

2.   A space of 0.30 - 0.33 m^2/animal without wool is sufficient. (0.25 - 0.45 m^2/animal had no influence on the loss rate).

3.   The strewing of superphosphate (100 kg/1 000 animals) is effective in elimination of high NH_3-concentrations. The superphosphate also inhibits the sporulation of *Cl. welchii* in the faeces, and so a source of enterotoxaemia infection is cut out.

4.   Protection against an endemic enterotoxaemia by vaccination with a toxoid (*Cl. welchii* type D) at the beginning of the journey can be effective, but vaccination with anacultures of *Cl. welchii* type D at the beginning of the passage is even more protective.

Plate 1.

Plate 2.

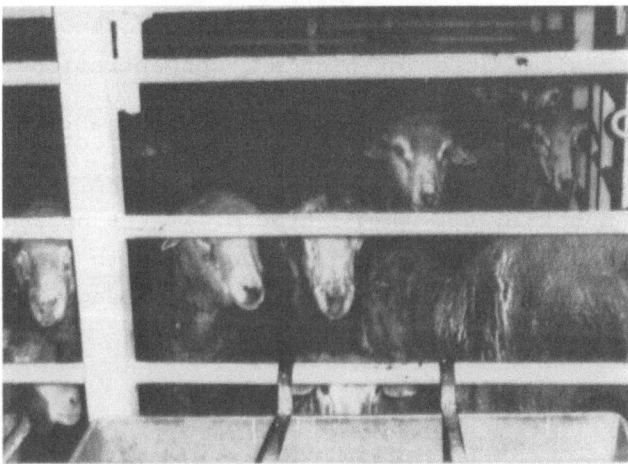

Plate 3.

DISCUSSION

G. van Putten *(The Netherlands)* How many animals did you have on the ship?

W. Müller *(Federal Republic of Germany)* We had 33 000 animals and our total losses were 1 421.

W. Sybesma *(The Netherlands)* Did these sheep suffer no heat stress?

W. Müller They did not. They had been sheared about four weeks beforehand. There was no heat stress. It was really *Clostridium welchii* infection. Those animals which we saw suffer from the disease died. There are a lot of questions to be answered here. We are working on the subject.

L.H. Huisman *(The Netherlands)* How many animals did you have per pen?

W. Müller It depended on the sheep. Sometimes there were 30 animals in the bows and sometimes 120 animals per pen. A reduction of the number of animals per pen would lead to better handling of the animals, but the animals are very quiet, especially sheep. The crew try to give them the best care, because they are interested. The main problem with transport is with the feedingstuffs used. They used pellets and the sheep come from grazing pasture.
As a result, typical stress appears in the rumen. Some of the literature says that there is too much protein, leading to infection, and others say that there is too much carbohydrate, but all the literature says there is a change. There is also in the literature the statement that you cannot cause infection just with *C. welchii*, there has to be a stress factor. In my opinion, the biggest stress is a result of the change in the feedstuffs. Sometimes they used a better pellet with more roughage, more hay, in it. These pellets lead to lower losses.

J. Melville *(Australia)* Unfortunately I do not have copies of the document, 'The Sea Transport of Sheep', published in March by the Australian Bureau of Animal Health. It is a report on the live sheep trade with the Middle East. It is, in part, critical of some shipping companies. The reason that I do not have any copies is the fact that it has caused a certain amount of controversy and it is already out of print. We are getting some extra copies printed, so I would hope to be able to supply it to anyone who is interested.
There are some comments which I think I should make. Firstly, we are exporting nearly 6 million sheep per year. The mortality rate overall per annum is marginally less than 2%. If you want to commute that, about half those animals would die on the farm if they were left behind, on the basis of the normal mortality rate of the sheep population of Australia.
Before I go into the report in detail, I should say that we have had several Government veterinarians on board ship and the report is the result of their findings. It is interesting, after hearing Dr. Müller speak, that on one vessel, which had 40 000 sheep, the major cause of mortality was acute lung pathology. Hyperacute lung pathology accounted for 17% of all deaths and acute lung pathology for 69% of all deaths. If my addition is correct, 86% of all mortality was due to pneumonia. The interesting thing about this is that in this particular vessel the younger sheep died first, during the initial stages of the voyage, and the older sheep seemed to die later. We started to get enteritis towards the end of this voyage.
On the question of ventilation, we have for many years published a set of guidelines for the construction of vessels and for ventilation rates. We have also had a set of guidelines on practices. One of the practices is preconditioning of sheep. They should be taken off pasture and held in a feedlot situation and brought onto pellets.
I do not think that heat stress is relevant, because what many Euro-

peans find difficult to understand is that many of these sheep come from temperatures which in summer can be in excess of 40°C. They are predominantly of
the Merino breed and are very well heat adapted.

In our guidelines, we only talk in terms of air changes per so many
minutes. This is worked out on the basis of deck space and the difference of
height between the decks. For example, if there is a clear 2.3 m between the
decks, the air there must be changed every three minutes. Ventilation capacity is worked out on the basis of the net volume of that deck space excluding
the sheep. The same is true for floor space. The figures that Dr. Müller
was quoting were very much in agreement with ours. They do vary. We worked
ours out on the basis of live animal weight. For a minimum total liveweight
of 20 kg we require 0.24 m^2 floorspace and for a liveweight of 80 kg we require 0.44 m^2.

The trend in sheep transport now is to go for above-deck carriage.
This clearly illustrates lower mortality rates for sheep carried above-deck.
We now have vessels which have up to eight decks above-deck, and there is
now one operating which carries 140 000 live sheep, and there is one under
conversion at the moment which will carry 180 000 live sheep.

The sort of problems which this report highlights are those which arise
when people step outside the guidelines. We have not had good enough control
on the vessels themselves, which means that they have not been inspected critically enough by marine surveyors before being allowed to load. That has been
changed. On the positive side, we are supplying some 300 000 breeding sheep
to Europe at the moment. We have one vessel doing this which carries about
30 000 per voyage. The last load which came off that vessel in which I was
involved lost only 75 animals. They were in-lamb ewes, so there was a positive increase in numbers, and samples showed an increase in bodyweight.

BULK TRANSPORT OF LIVESTOCK BY SEA WITH PARTICULAR REFERENCE TO INSTRUCTION, PROCEDURE AND PROBLEMS

C. Platt

Regional Director,
World Society for the Protection of Animals,
106 Jermyn Street, London SW1, UK.

INTRODUCTION

In spite of rapid advancements in recent years in the bulk carriage of livestock, notably in the design of vessels, equipment, between deck facilities, improved nutrition and the elimination of some stress inducing factors, inherent problems still exist. To maintain progress, these problems must be first identified and discussed so that remedial action can be taken. This concept is undoubtedly in the minds of the organisers of this seminar.

The emergence of vessels specially designed or adapted for the bulk carriage of livestock has been the most significant advancement and this paper is addressed to this type of vessel. It is, however, relevant that the greatest number of vessels carrying livestock in the areas with which I have experience - the Far East, Africa, Red Sea, Mediterranean and the Aegean - are not bulk carriers by design, only by the numbers of animals they embark amongst other cargo. Until reciprocal international standards are set for carriage of animals by sea, to those carried by air transport, I can see no practical means of regulating the widely diverse conditions that apply to these general cargo/livestock vessels.

With the emergence of the custom built bulk carrier, certain procedures, if not internationally mandatory standards, have evolved to facilitate the loading and transit of large consignments of livestock. Problems have also evolved and whilst each vessel has its characteristic problems, it is those that apply generally on which I will concentrate and notably those with which I have had first hand experience.

PRE-LOADING

Environmental adjustment

Experience shows that stock become more manageable and subsequent loading is less stressful following a period of conditioning at assembly lairages close to embarkation point. This period of adjustment is particularly relevant to stock from free range habitats, unaccustomed to the alien conditions created by dockside and shipboard environments.

Diet

For reasons of conserving space in the feed holds of large bulk carriers, shipboard diet comprises a high proportion of pelleted concentrates which vary considerably in quality and ingredients. The adjustment period at assembly lairages enables stock to be gradually eased onto a shipboard diet, thereby reducing the problem that sometimes arises of stock initially rejecting unaccustomed feed after embarkation. As an example, some exporters assemble stock at least seven days prior to embarkation and commence with a diet of 75% hay to 25% pellets, gradually reversing the proportion as embarkation approaches. Needless to say, it is advantageous to use the same pelleted feed in the assembly lairages to that used on board.

'Topping-up'

It is common practice to 'top-up' consignments to make up the numbers of those rejected during the final pre-loading check, or for other reasons. Sound reasons exist to have a reserve of pre-conditioned stock but this policy is not always applied and unconditioned stock are brought from nearby farms or holding lairages. Analyses of the stomach contents following *post mortem* examinations of early shipboard casualties show that a high proportion originate from these unconditioned 'topping-up' stock.

Casualties

Contrary to national legislation requirements, casualties necessitating immediate slaughter are left for unreasonable periods for the following reasons:

a) reluctance on the part of the relevant authority to order the destruction of casualties.

b) conflict between the buyer's agent and the authorities over the need for immediate destruction.

c) long delays before local butchers arrive to negotiate purchase of casualties.

A need exists for more rigid implementation of this legislation.

While on this subject I would draw your attention to avoidable injuries caused by stockmen failing to remove binding wire and twine from baled feed carried into the lairages. Lacerations of the mouth, foreface and legs have been identified and, in extreme cases, entangled twine has restricted blood circulation, resulting in lameness.

LOADING

Obstructions

The pen gates on some bulk carriers are lifted and lowered by means of cables which can become entangled in the horns of cattle and sheep. This impedes the flow of animals along the passageways and can cause crushing and injuries. Designers should be advised to examine alternative lifting and lowering devices or re-position the cables.

Allocation of pen space

Due to the behaviour of sheep - their tendency to bunch together and their speed of movement when compared with cattle and horses - it is difficult for stockmen to get the correct numbers in the pens on the animal decks. There is divided opinion as to whether these animals should be moved when loading is completed and before the vessel sails, or to allow them to settle and remove them at a later time.

I suggest that any adjustments should be made before
sailing because in the event of initial rough weather conditions,
injuries could be caused in both overcrowded pens and partially
occupied pens.

DURING VOYAGE

Stowage of feed

Many bulk carriers stow all feed in one hold of the
vessel. In my view it would be more practical to separate
the feed into different holds to prevent all being destroyed or
spoiled in the event of fire or water seepage occurring in mid
ocean.

Feed and water receptacles

With bulk sheep carriers pen space is designed to accommodate about 50 - 60 head. Each pen contains three feed and
three water receptacles, usually constructed of opaque plastic
to facilitate cleaning which is done by pressure hose. The daily
ration of feed is usually about 1 kg/head in two feeds, each
trough holding about 22 kg. Continuous water is supplied and.
controlled by ball and valve mechanism in the water receptacles.
The problem of contamination by feed, faeces and urine has yet
to be overcome. Identification of heavily contaminated receptacles servicing second tier lairages, could be more easily
recognised if transparent plastic was used instead of opaque.

Livestock supervision

In some cases this is inadequate. Several bulk carriers
are in service carrying 90 000 - 100 000 sheep on six decks
with about 5 - 6 stockmen per deck. Supervision is usually
carried out by one livestock officer who is responsible for
nutrition, health, treatments, destructions, records and the
monitoring of temperature and ventilation levels. To reduce
the current losses of stock, averaging at 1.2%, serious consideration should be given to one livestock officer and one veterinary surgeon accompanying such consignments as a mandatory
requirement by the exporting authority.

Ventilation

Normal air flow through the animal decks' ventilation systems has been measured at 0.5 m/sec at the source which is barely adequate to dissipate totally excessive levels of humidity and ammonia in the centre pigs. This causes problems at night when stock are resting. High concentrations have been metered at 0600 h but subsequent readings taken at 0900 h showed that the movement of the stock during feeding had caused the build-up to disperse. The installation of fans over the centre pens is indicated to disperse build-up of fumes when stock are resting. Further tests into tolerance levels indicated no apparent ill effects when stock were subjected to ammonia levels of 10 parts/million, but after 5 min exposure to levels of 20 parts/million, irritation to the nose and eyes were observed. The installation of a warning system to monitor air flow appears a desirable precautionary measure.

Sickness and mortalities

Losses of 1.2 - 1.4%, which are not uncommon, represent 1 000 - 1 500 head of stock on the larger sheep carriers. With voyages of 10 - 14 days duration, there is a significant loss of condition with accompanying sickness and mortalities during the latter half of the voyage. Most common are diseases of respiratory and gastro-intestinal tracts; also infectious kerititis (pink eye).

The build up of ammonia and humidity in the atmosphere, as previously mentioned, plus the large amounts of dust particles from the pelleted feed, are contributory factors and remedial action is possible. However, it would be unwise to ignore the harmful effects of progressive stress when subjecting livestock to these environmental conditions for long periods.

To counter this loss of condition during the latter stages of the voyage, rest periods during normal hours of darkness are reduced to encourage a longer feeding period. This, in my view, merely adds further stress to already possibly over-stressed animals and this practice should be discouraged.

Deaths amongst sub-adult animals on vessels using mixed grade penning gives rise to concern, as carcase examinations indicate that these animals have lost out in their competition for nutrition with older and more aggressive animals. A more judicious grading system at the time of loading to separate immature stock could avoid this problem.

Destruction of casualties

Provision for a captive bolt pistol or impact stunner on board bulk carriers is desirable to facilitate destructions with the minimum disturbance to the stock.

UNLOADING AND SUBSEQUENT DISPERSAL OF STOCK

This is a most critical period following a long sea voyage and, clearly extreme care must be exercised during unloadings. Numerous areas of concern have been identified but these would not fall within the terms of reference of this seminar.

I will conclude by drawing your attention to the need to rest stock for at least 24 h following disembarkation before further transit is attempted. I make this recommendation in the light of witnessing animals go down when consigned to rail and road transport immediately following disembarkation. Adjustment to new environmental conditions is most important but frequently ignored with disastrous consequences.

DISCUSSION

G. von Mickwitz *(Federal Republic of Germany)* We worked for some years with cattle being transported to Africa by air. The result of this was that in Germany we now have some recommendations as to the minimum requirements for the transport of cattle by sea and the transport of animals by air. I think these were published by the Ministry of Agriculture, Forestry and Food a year ago. The transport of animals by aircraft is very comfortable for the animal. When I transported calves by air, if there was enough room, I found that within ten minutes all the calves lay down and slept until we arrived. When you transport cattle by air, it is not possible, if the aircraft is full, to allow them so much space that they lie down. You have to tie them tail/head/tail/head and so on. When you have 75-80 cattle in one aircraft you have to crawl over their backs to control them. If the duration of the flight is only five hours, such as a flight to Tunis, perhaps, this is a good system, as there is no danger to the cattle, even to pregnant cows. What about flights of a duration of 24 h, such as flights to Brazil? How can we force cattle, and particularly pregnant cattle, to stand without being able to lie down?

If you have good transport conditions on a ship, and you have enough space, say, 1.80 m^2 for each animal so that they can lie down, then the animals will only suffer from seasickness for one or two days and then their behaviour will return to normal. So that is my problem: what is the limit to the time for which we can force pregnant cattle to stand without lying down?

M.E.T. Watts *(UK)* I think that if one observes cattle on any form of transportation they want to lie down approximately 4 - 6 h after the start of the journey. If you have a pen of eight animals, then about 25%, i.e. two or three, will lie and after a period of a couple of hours another two or three will lie down and the original animals will stand up. I think we would find, if we observed pregnant animals, that after a period of 6 - 10 h one should leave space available for every animal to lie.

J.E. Melville *(Australia)* We are airfreighting animals to Australia from this part of the world. One of our major concerns, having made the decision to permit import, is to maintain the animal's health status. Consequently they have to fly 'west about', in other words they have to fly through Canada rather than the direct route through the Middle East. Our problems still come back to the same sort of problems that we are getting with our live sheep shipments. There are requirements, there are guidelines and in the main, where we have trouble these requirements have not been met. The single biggest problem we have with flying animals that distance, and I am talking about 36 h, is whether air-conditioning equipment is available on the ground at refuelling stops. We have insisted on this requirement, and we have, in fact, refused to issue permits to some airlines because they have been negligent in the past.

A. Cuthbertson *(UK)* How great are the difficulties in encouraging the importing countries to import animals for slaughter not in the live form but as carcases? I know this is done, so why is it not done more widely? Why do we have to ship these large numbers of lambs from Australia to the Middle East? Will the importing countries not accept the lambs in carcase form?

J.E. Melville There are two simple answers. One is religious, but the main reason is that they have no chilled or frozen meat distribution network. By tradition in the Middle East and in South-East Asia, people do not eat meat unless it is 'jumping hot'. It goes straight from slaughter to a retail outlet and then straight to the consumer.

P.V. Tarrant *(Ireland)* There were traditional and economic reasons for live animal importation. For example, the importation of Irish cattle into Britain is not done for religious reasons or for reasons of lack of refrigeration or the desire of the British to eat 'hot' meat. It is done for traditional reasons and economic reasons associated with farming patterns which result in the continuation of the live trade in meat animals.

P.J. O'Connor *(Ireland)* I have been most interested in all papers, and particularly in the last one, in reference to discharge. I think there is a need to make a priority for livestock at ports of arrival. We have reports from Ireland where shipments are held up for two or three days at anchorage outside the port. This cannot be beneficial. I would like to see a worldwide recommendation that livestock should get a priority and that berthage should be available for livestock at all times. Indeed I would expect services to be 24 h services where animals are concerned.

What about stability of vehicles on roll on/roll off ferries because this is at risk on a sea voyage? Is it UK policy that these vehicles would not go on RO/RO voyages of more than 12 h duration?

<u>M.E.T. Watts</u> It is normal sea practice in cargo holds which carry these lorries to hold the lorries down to the deck by use of chains and shackles. These are very strong, with a breaking strain of about 20 t and they are located every 2 - 3 m along the side of the vehicle. Originally the point of attachment to the vehicle was the chassis, but these vehicles are quite often detachable or they are being towed by a different towing unit, and we now specify that they should be held with shackling units on the container itself. Stability is also significant in the two-tier type of vehicle, particularly if they are carrying adult cattle, where you have quite a heavy weight 3 - 4 m above the ship's deck. Under these circumstances we would expect some shackling to take place on the upper deck of the vehicle.

As far as the duration of the journey is concerned, provided the environment can be controlled for all stages and provided that there is adequate space for the animals to lie and behave as naturally as possible, and provided that there are facilities for providing food, water and other attention, the duration of the journey itself is not critical. It is no different from any other length of journey. In the transcontinental trade there is often a 3 - 4 day journey, from England to Portugal, for example, so there is no reason why a ship's journey should not be similar in length as long as the facilities are provided.

<u>G. von Mickwitz</u> In Europe we have a number of large shipping companies which carry livestock. All have good equipment and try to handle animals well.

However, we have sometimes found insufficient light is provided to be able to inspect animals properly, and loading densities can be exceeded. We have certain minimal requirements and have found, for example, if there is less than 5.4 m^2 for pregnant cattle the percentage of abortions increases.

It is important to have uniform regulations internationally.

<u>G. van Putten</u> *(The Netherlands)* I know that there are very detailed regulations in Germany and we have been working hard on draft regulations in the Netherlands. There are very good regulations in Switzerland and in France. I do not know of any detailed draft regulations being prepared in the United Kingdom. Am I wrong there? I would like to know if there are any draft detailed regulations in the UK?

<u>R. Moss</u> *(UK)* I am not quite sure which regulations Dr. van Putten is considering here. I believe that what he is thinking about are the regulations which may follow Directive 81/389. In the UK we have, since the middle of the nineteenth century, had very detailed regulations relating to the carriage of livestock by sea, by canal, by road and, in the last few years after we ratified the international convention in relation to transport, by air. The problem, I think, of detailed regulations at the present time, and this may come out in the discussion this afternoon, is that we have made certain regulations - we have decided on the square footage for particular apertures and so on - without really knowing whether animal behaviour or physiology fits. We have done this on what I would call a hunch, or on the basis that it has always been done that way. Dr. van Putten himself said that there should be a particular slope. There are many ramps and many regulations throughout Europe which do not follow the point that Dr. van Putten was making. Ramps are of quite different sizes. We must be very, very careful in moving into detailed legislation which cannot be backed up either by practical application and knowledge or scientific fact.

<u>G. von Mickwitz</u> I agree that you must have scientific fact. The main point is always the loading density, because that means money.

SESSION V

PRESENT RESEARCH BEING UNDERTAKEN
AND CONSIDERATION OF WHAT FURTHER STUDIES, IF ANY,
WOULD BE DESIRABLE

Chairman: P.V. Tarrant

TRANSPORT OF ANIMALS INTENDED FOR BREEDING PRODUCTION AND SLAUGHTER

D.B. Stephens

ARC Institute of Animal Physiology,
Babraham, Cambridge CB2 4AT, UK.

ABSTRACT

It is usually necessary to transport farm animals at least once during their lives. Some animals are transported within the first 2 - 3 weeks of life, often over long distances. As the animals grow and progress via their various stages of development it is often necessary to transport them from one specialised unit to another. All meat animals are finally transported to the abattoir for slaughter. The recent developments in superovulation and embryo transplantation techniques have focused attention on the possible effects of transportation on the ovarian response to gonado-trophic stimulation in the donors and on embryonic survival in the recipients.

The experiences associated with transportation are usually novel to the animal and therefore produce an acute unconditioned emotional stimulus. The physiological and psychological responses induced in the animal have provided the basis of several recent reports on the effects of transport-ation on farm animals.

Under field conditions, the mortality rate in calves that are purchased and transported at an early age can be as high as 23% (Staples and Haugse, 1974). Calves that die after a journey are usually dehydrated and often have pathological changes in the kidney. This had led to the develop-ment of techniques to study the effects of transportation on cardiovascular parameters and renal physiology in particular.

Several studies have reported the various effects of transportation on reproductive performance. In our laboratory, real or simulated transport did not produce an effect on the age at which puberty was attained in the gilt. Simulated transport had no effect on pregnancy, parturition or viability of the newborn, in the rabbit.

A major factor which contributes to mortality in animals which are transported at a very young age is dehydration, usually resulting from the fluid loss which accompanies scours. It is suggested that future studies be directed along lines which will produce information on:-

a)   mechanisms involved in controlling fluid and electrolyte balance in the very young animal

b)   how such mechanisms change and develop with age, particularly in relation to their possible modification as a result of exposure to potentially adverse situations such as transportation

c)   the role of transportation in the aetiology of scours.

Finally, it is suggested that a combination of (i) ethological observations of animals under real transport conditions at progressive stages of journeys of various durations be obtained plus (ii) physiological and behavioural observations on animals during real journeys and in the laboratory under simulated transport conditions. Such information is essen-tial in order that we may evaluate the welfare problems of farm animals during transportation.

GENERAL INTRODUCTION

Domestic animals which are kept for breeding, meat or
milk production are normally transported on only a few occasions
during their lives. Different stages of the life of farm animals
are usually spent on specialised units. For example, pig rearing
farms and hatcheries are often separate from the final fattening
units. The animals are therefore transported from farm to farm -
sometimes over long distances. The reason for the movement is
usually due to the geographical variations between regions. For
example, grass which grows abundantly and cheaply in the Western
Atlantic coastal regions of Europe, results in the predominant
dependence of farmers there on dairy enterprises which means
that there is a surplus of calves. The male calves, particularly
the Friesian, are suitable for fattening into beef on cereal
based diets, usually in the cereal growing eastern regions of
Britain. In addition, the demand for veal in Continental Europe
has led to the use of breeds of sire such as the Charolais. When
crossed with the British dairy breeds, such sires produce progeny
which are particularly suitable for veal production. This breeding practice has led to an increase in the number of calves that
are transported from Britain to Europe. The age of the calves
at transportation normally varies between a few days and one
month, depending on the husbandry methods practised on the farm
of origin.

Mature animals are often transported for breeding purposes,
particularly in connection with the recently developed techniques
of embryo collection and transplantation from genetically
superior ewes, cows or sows. This technique is normally carried
out at specially equipped premises.

PHYSIOLOGICAL AND BEHAVIOURAL STUDIES UNDER FIELD AND LABORATORY
CONDITIONS IN ANIMALS DURING TRANSPORTATION

Introduction

Knowledge of the ability of animals to cope with journeys
of various length by land, sea or air is important, particularly
from the welfare standpoint. The procedures for monitoring the
physiological and behavioural responses of animals during

transportation have developed along two main lines. Firstly, behavioural observations have been made on animals during the course of normal journeys (details are extensively reported elsewhere in this volume). Secondly, situations have been created in the laboratory or the field where it was intended to simulate conditions which exist during normal journeys.

Physiological considerations

Adverse stimuli associated with transportation can be divided into two groups:-

a)   Physical disturbances. Examples of these are:

   i)   rapid changes in weather conditions
   ii) physical agitation due to motion of the vehicle
   iii) inadequate lighting
   iv) unsuitable floor material for walking, standing or
        lying in comfort.

b)   Emotional disturbances. Examples of these are:

   i)   breaking of the bond with the dam on leaving the
        farm of origin. This is potentially very disturbing
        and often overlooked
   ii) exposure to stimuli which produce excessive fear
   iii) the establishment and maintenance of new social
        relationships.

The responses elicited by adverse stimuli (stressors), such as the examples given above, are normally behavioural or neuroendocrine in nature. The classic physiological response of animals to alarm were originally described by Cannon (1929) and later elaborated by Selye (1950).

Fear is usually listed among the emotions and can trigger the physiological responses of alarm (Gray, 1971). Clearly, each new phase of a journey is novel to the animal and therefore potentially fearful. Hence the psychological components of adverse situations should be considered in parallel to the more familiarly understood physiological changes which often occur rapidly during the course of transportation.

Field observations

Mortality

Transportation can result in a high mortality rate, especially when animals are transported at an early age. For example, calves which are transported over long distances when below 14 days of age have shown a mortality rate as high as 23% (Staples and Haugse, 1974). Transportation is often associated with problems of scouring in calves which often leads to dehydration. Macroscopic observations of kidneys of calves which died in transit usually revealed softness to the touch which indicates some degree of tissue degeneration.

Blood urea levels in transported calves

Transportation has been shown to result in a decreased plasma concentration of glucose and total proteins with a concomitant increased level of urea and uric acid (Agnes and Genchi, 1972). These authors suggested that such changes resulted from the muscular activity and starvation imposed by the journey. Observations on the effect of a road journey on kidney function in calves have been carried out in a preliminary field trial in our own studies. In these experiments, the calves observed were intended for slaughter at a very early age (usually under 1 week) (called 'Bobby Calves' in Britain). Such calves are usually young and therefore small (30 - 40 kg body weight approximately).

A group of fourteen such calves was made available at a market (Carmarthen) for blood sampling. After the blood samples were taken for urea estimations, the calves were loaded onto a standard cattle transporter capable of carrying up to 100 calves. During the journey, the experimental group was kept together by being penned off from others being transported. After the journey (of approximately 4 h) to the abattoir (Nantwich), each individual calf earmarked for observation was unloaded, and a second blood sample was immediately taken for a second blood urea analysis.

Results

The effects of transportation on blood urea levels are shown in Table 1. There were no statistically significant changes.

TABLE 1

TRANSPORTATION EFFECTS ON BLOOD UREA IN CALVES
BLOOD UREA (mg/100 ml; N = 14)

	Mean	SEM	Range
Start (Carmarthen)	50.2	± 5.6	29.2 - 113.6
End (Nantwich)	59.1	± 7.5	33.0 - 147.0

Mean ± SEM values of blood urea concentrations for a group of 14 calves at the beginning and end of a 4 h road journey

Laboratory experiments

Introduction

It has been evident from studies on several laboratory species (Pappenheimer, 1960; DiSalvo and Fell, 1971; Mancia et al., 1974) that situations which are known to evoke emotional responses can profoundly influence blood flow to and within peripheral organs, particularly the kidney. Renal vessels are richly supplied with sympathetic nerve fibres and direct peripheral stimulation of such fibres produces a decrease in renal blood flow (Fiegl et al., 1964). The role of the central nervous system in the control of the renal circulation is not entirely clear (Mancia et al., 1974) although intense renal constriction can be induced by hypothalamic stimulation (Fiegl et al., 1964). When a state of emotional arousal is sustained in the rat, it has been shown that chronic interstitial nephritis can develop (Henry et al., 1973).

Since it appears that there is a degree of emotional arousal in domestic animals exposed to the novelty and other factors associated with transportation, we have begun a series of experiments designed to monitor fluctuations in renal blood flow in relation to simulated and real transport. Another objective was to test the hypothesis that fluctuations in blood

flow to the kidney provide a reliable and sensitive indication
of small and otherwise undetectable psychological responses.
From a clinical view point, even small redistributions of
electrolytes and concomitant fluid imbalance can result in
serious effects, such as scouring, in neonatal animals (Bywater,
1977; Bywater and Woode, 1980; Bywater, 1980).

Effects of simulated and real transport on renal blood flow in the calf

As a first step, experiments were designed in our laboratory to monitor fluctuations in blood flow in the renal artery
of calves and pigs. The equipment employed consisted of an
ultrasonic flow meter (Rader and Stevens, 1974) which was placed
surgically around the renal artery. To ensure that the physical
conditions could be accurately reproduced, a Transport Simulator
(TS) was constructed (Figure 1). It consisted of a wooden, steel
framed structure attached to a fixed base. The base consisted
of a steel plate which rode on large ball bearing structures
which in turn were attached to a cam follower. As the wheels of
the cam turned, the two sides of the box moved up and down a
distance of approximately 14 cm, independently of each other, at
a frequency of 30 movements/min. This type of motion was also
accompanied by a slight lateral movement of approximately 11 cm
at a frequency of 35 movements/min.

Results

There was a small decrease, of approximately 5% in renal
blood flow during the first few minutes of exposure to the TS
compared with control levels in the undisturbed animal. After
a few minutes, the renal blood flow regained normal levels and
was then elevated for the duration (45 min) of transport
simulation. Heart rate (HR) increased from below 100 beats/min
to over 140 beats/min and remained at this level for the duration of the transport simulation. In the pigs, under identical
conditions, the results were essentially similar.

Fig. 1. Transport Simulator in laboratory with entry ramp.

The effects of transportation on heart rate and plasma corticosteroid levels in the calf

In another experiment (Stephens and Toner, 1974; Stephens and Toner, 1975; Toner and Stephens, 1975), HR was continually monitored during normal transportation by livestock vehicle in calves aged 2 - 4 months. In this experiment, with the calf standing and unfettered, HR was elevated from a resting level of about 80 beats per minute (b.p.m.) to about 110 - 115 b.p.m. for 1 h while the transporter was in motion. When the calf lay down the HR was reduced to near 80 b.p.m., even with the lorry moving. This reduction in heart rate which accompanied lying down was probably due to both psychological and physiological influences. The redistribution of blood between organs resulting from the change in posture affects HR (Dauncey and James, 1979). In the human (Dauncey and James, 1979) it was shown that there was a marked increase in HR when the subject stood up.

This increase in HR was greater when the subject stood still than when small movements of the limbs were made. In another parallel experiment, a calf was placed inside a wooden cubicle, which was then placed in the lorry for the duration of the journey (1 h). The size of the cubicle was such that the calf could lie down but not turn. Although the increase in HR which occurred during the journey in this situation was similar to that found in a calf carried free, the increase in the plasma corticosteroid concentrations was less.

For example, in one calf, the basal corticosteroid concentration in plasma was 6.9 ng/ml (Stephens and Toner, 1975). When transported unfettered in the lorry, the level had increased to 14.3 ng/ml after 30 min of transporting and, decreased to 8.1 ng/ml 45 min after being returned to the loose box. The corresponding values for the same animal transported while housed in a wooden cubicle which was placed in the lorry, were 8.4 ng/ml in the resting state. After 30 min of the journey the concentration had only raised to 9.5 ng/ml and had returned to 6.9 ng/ml 45 min after completion of journey.

It was not clear why there was an attenuation of response whilst inside the cubicle. The sensory input of the calf was obviously reduced by its being placed in a restricted environment and also the direct effects of the vibration imposed by the transporter were probably reduced because the calf was firmly supported by the sides of the cubicle.

The effects of simulated transport on digestive function in the calf

The effects of transport simulation on digestive function in the calf have been investigated in our laboratory. The dry matter (DM) content of samples of faeces were used as an index of whether the treatment produced scouring in the calf. In addition, the effects of the time of feeding in relation to the disturbance were examined by comparing the effects of feeding the calves before and after transport simulation. These experiments were repeated using milk substitute powder at twice the recommended concentration, since it has been claimed that 'digestive scours' can develop if calves are overfed immediately after a journey.

Five male Jersey calves, aged 1 - 3 weeks from the
Institute herd were used over a period of approximately a year.
The calves were examined clinically each morning to ascertain
normal health. The protocol of the experimental observations
are summarised in Table 2. The order in which the treatments
were carried out was similar in all experiments, as indicated
in Table 2. Therefore, the possibility of habituation must be
considered. A normal feed consisted of 227 g of calf milk
substitute made up to 2 litres with tap water and fed at $37^{o}C$
via a rubber teat.

TABLE 2

PROTOCOL OF THE EXPERIMENT

Experiment Number	1	2	3	4
Meal concentrations	Normal (227 g/2 l)	(As 1)	Twice normal (500 g/2 l)	(As 3)
Time of feeding in relation to transport simulation	1 h before start	1 h after end	2 h before start	1 h after end

The calves were taken by trolley over a distance of 500 m
to the laboratory and exposed to 2 h of transport simulation.
On the following day, faeces samples were collected for DM
analysis by fitting specially designed collection harnesses to
the calves for approximately 6 h. Usually a 2 - 3 day interval
was allowed between each experiment. Control faeces samples
were collected on the day before the experimental series.

Results

The exposure of Jersey calves to short periods of simulated transport had no effect on the DM content of the faeces,
as shown in Table 3. Overfeeding, either alone or in combination with transport simulation, also produced no detectable
effect. All 5 calves appeared clinically normal throughout
their entire sojourn in the laboratory (about 1 month).

TABLE 3

EFFECTS OF SIMULATED TRANSPORT ON MILK DIGESTIBILITY IN CALVES

	Meal size	Simulated transport (2 h)		Control	
		Mean (% DM)	Range (% DM)	Mean (% DM)	Range (% DM)
A Fed 1 h pre exposure	Normal	18.7	16.4-30.1	22.7	14.9-35.2
	Over-fed (2 x normal)	20.2	11.4-39.4	14.8	16.2-33.8
B Fed 1 h post exposure	Normal	16.3	11.1-30.4	24.7	12.4-29.8
	Over-fed (2 x normal)	19.5	9.6-36.8	17.3	8.8-40.3

Dry matter (DM) content expressed as percentage (%) of samples of faeces collected from calves exposed to transport simulation and/or overfeeding. See text for details.

EFFECTS OF SIMULATED TRANSPORT ON HEART RATE, BLOOD PRESSURE AND RENAL ARTERIAL BLOOD FLOW IN PIGS

The techniques for monitoring cardiovascular responses to real and simulated transport in calves have also been applied to pigs aged 4 - 5 months in our laboratory (Stephens and Rader, 1981). The resting control level for HR in our pigs was around 100 - 110 b.p.m. Although HR was elevated to over 150 b.p.m. in all pigs during the first half hour of exposure to the TS, there then followed a gradual reduction of HR to near normal resting levels. After several exposures to the TS there was a less marked increase in HR during transport simulation, presumably due to the process of habituation. No significant changes in blood pressure (BP) were detected during the experimental manipulation. Following a transient decrease in renal blood flow at the moment when the simulator was switched on, renal blood flow increased to about 20% above the resting control value.

Measurements of HR, renal blood flow and BP were also carried out at the beginning and at the end of a 4 h road journey. The pig was carried alone and free standing. In this case there was a significant decrease in the renal blood flow while the animal was eating prior to the journey. There was also a slight decrease in renal blood flow and an increase in HR immediately on entering the lorry. However, the renal blood flow quickly regained normal levels even before the start of the journey. At the end of the journey, the HR had returned to near the control rate, but the renal blood flow was slightly elevated.

Cardiovascular changes were also monitored in pigs exposed to transport simulation when dehydrated following a 48 h period of water withdrawal. In this state, the HR remained elevated for the whole duration of exposure to the TS. The initial decrease in renal blood flow at the moment when the TS was switched on was more marked than that seen in normally hydrated pigs. It was of interest that this decrease in renal blood flow persisted throughout the entire exposure to simulated transportation, compared with the increase which was observed in normally hydrated pigs.

EFFECT OF TRANSPORTATION ON REPRODUCTION

The attainment of puberty

Incidental observations by veterinarians and pig farmers on pre-pubertal gilts have led to claims that various husbandry practices can induce puberty in the gilt. Although it has been suggested from studies on gilts reared in isolation (Signoret, 1970) that there is an innate pattern of sexual behaviour in the gilt, several studies exist which suggest that attainment of puberty can be modified by environmental conditions.

Transportation is often implicated as a factor that can contribute to timing of first oestrus. Groups of gilts are regularly brought from various farms in East Anglia to the Institute to provide material for experimental purposes. It has been very commonly observed that a large proportion of such gilts often exhibit oestrus within a few days of arrival. Clearly, practical methods which could be applied to induce the early attainment of puberty would have productive and economic advantages. For these reasons, controlled experiments have been carried out on pre-pubertal gilts to see if various management practices can influence the onset of puberty in the gilt (Close et al., 1981).

Results

Gilts exposed to real and simulated transport for periods of 1 - 2 h showed oestrus at between 218 - 224 days of age, while control animals left undisturbed, showed oestrus within the same age range. Exposure of similarly aged gilts to simulated transport plus artificial boar odours produced by aerosol ('Boarmate', Antect International, Surrey) again showed no effect. The age of puberty in these groups was attained at a mean of 218 days. However, gilts moved to the vicinity of sexually mature breeding males and females attained puberty significantly earlier at a mean age of 181 days. Therefore, it appears that, as shown in the rat by Mandl and Zuckerman (1952), vibration and other stimuli associated with transport have no influence on the onset of puberty. Apparently the main stimulus for attainment of puberty, at least in our experiments, was the proximity of adult breeding stock.

Fertility and pregnancy

Other important reproductive criteria such as time of ovulation, fertility and the course of pregnancy are known to be affected by various kinds of emotional and environmental disturbances in many mammalian species. Since transportation is believed to be capable of producing changes in the hypothalamus-pituitary-adrenal axis, it is possible that it also plays an important role in producing functional changes in the ovaries and reproductive tract. The oestrous cycle in the ewe is affected by transportation (Averill, 1964) as also is the timing of ovulation (Moberg, 1976). Anoestrous ewes will ovulate within a week of the commencement of a prolonged journey (Branden and Houle, 1964; Lang, 1964). The relatively mild disturbance associated with transference of ewes to a strange field within the first 20 day period *post coitum* has also been shown to result in a significantly higher rate of embryonic loss compared with the losses found in normal undisturbed controls (Doney et al., 1976). Housing sheep for a short period prior to transportation minimises the cortico-steroid elevation produced as a result of the disturbance

(Reid and Mills, 1962), as does transporting them with lambs at foot (Thurley and McNatty, 1973).

In the cow, transportation has also been shown to affect the time of ovulation and fertility. For example, Short-Horn heifers have been shown to ovulate soon after the end of a journey (Lamond, 1962). Artificial insemination of adult cattle shortly after transportation to new farm holdings invariably results in a significant reduction in conception at first insemination (Holt, 1962).

The possible relevance of transportation effects on application of embryo transplant breeding techniques in farm animals

New developments in animal breeding methods and techniques such as the procedures associated with superovulation, collection and transfer of fertilised ova from valuable donors into recip- ients, often necessitates the transportation of such breeding stock to and from specialised units established specifically for the purpose. Therefore, the influence, and possible deleterious effects, of the disturbances associated with transportation on fertility, pregnancy and viability of the neonate assume consider- able practical importance. However, due to the high cost of using farm animals experimentally, rabbit does have been used in our laboratory to test their feasibility and usefulness as models for studying the effect of transportation on fertility and the course of pregnancy. In support of the choice of the rabbit was an observation by Lee and Ring (1956), who found that administration of ACTH resulted in a high level of foetal mort- ality, followed by a lowering of mean birth weights and overall viability of the newborn. Direct administration of hydro- cortisone in the rabbit at various stages of pregnancy can also lead to foetal resorption, small birth weights and reduction in viability of the newborn (Courrier and Cologne, 1951; Robson and Sharaf, 1952).

Observations on the effect of simulated transport on pregnancy, parturition and respiration in the rabbit

In experiments at our Institute (Adams and Stephens, 1980),

mated does were transported by road over a distance of 9 miles
and then exposed to either:

A)   being left undisturbed in a quiet room, or

B)   being left on the floor near the TS to be exposed to the
      loud noise only, or

C)   placed inside the TS so as to be subjected to a combin-
      ation of physical agitation plus the loud noise.

Control does, which were also pregnant, were left undisturbed
in their home cages. In order that the degree of emotional
arousal produced by the treatments could be estimated, the
respiratory rates were monitored in separate experimental groups
of rabbits (male or female). They were treated to: (a) movement, (b) noise or (c) noise plus agitation in a similar way to
the pregnant does.

Results

Of 49 rabbits which were transported by road, 47 maintained pregnancy to term. This number compared with 11 out of
12 does which maintained pregnancy to term in the undisturbed
group left as controls. The length of gestation, number of
stillborn and litter size was not affected in any way by the
experimental treatments designed to simulate transportation.
There were therefore no apparent effects on reproductive
performance, even though the rabbits were clearly aroused and
disturbed by the exposure to the noise and agitation of the TS
as was made evident by the marked increase in respiration rate.
It appears that the rabbit, at least under our experimental
conditions, is resistant to the effects of short term exposure
to real or simulated transport. Therefore, the rabbit does not
appear to be an appropriate subject to provide a model for
investigating the effects of transportation on pregnancy in
ewes, cows or sows.

CONSIDERATIONS OF FURTHER EXPERIMENTAL STUDIES

      Ideally, the conditions under which farm animals are
transported should provide optimum conditions which satisfy the
health and production aspects of animal management and also the
basic welfare requirements of the animals. In reality, it is

probable that ideal husbandry practices also provide the best
productivity and welfare conditions. Research aimed at elucidating the optimal conditions for the transport of farm animals
could therefore proceed along considerations of (a) productivity
and (b) welfare requirements.

Productivity

One of the major causes of weight loss and even death in
young animals such as calves and piglets is dehydration.
Clinically, dehydration is said to result from the digestive
upsets and scours which, incidentally, are often seen in calves
that have been transported (Bywater, 1980). Clearly, therefore
an increase in the knowledge of the basic mechanisms which (a)
cause digestive upsets and (b) control fluid and electrolyte
redistribution within the young animal, particularly when
exposed to a stressor such as transportation, would be highly
desirable. The mechanisms which control the distribution of
water and electrolytes at the neonatal stage are probably less
developed and therefore less competent than that of an adult.
Knowledge of factors affecting the ontogeny and time scale
during which the animal attains competence in maintaining fluid
and electrolyte homeostasis in the face of adverse situations
would help in making decisions as to (a) minimal age for transportation, (b) the duration of each journey and (c) the optimal
times for offering fluids and electrolytes to the animals in
relation to rest periods.

The question of whether the animals should be transported
within single cubicles or in groups as well as the optimal size
of such groups is also important. For example, it may be that
isolation of a calf during transportation produces a larger
endocrinological disturbance than that produced when kept in a
group. Any extreme endocrinological changes could directly
produce a detrimental effect on fluid balance leading to a
lowered resistance to pathological challenge.

Welfare considerations

It is probable that the basic welfare considerations
affecting animals during transportation are closely allied to

those concerned with health and productivity, as described in
section A above. When discussing welfare considerations, the
psychological responses must be taken into account as well as
the more readily detectable physiological and thermal responses.
However, the experimental study of psychological responses and
the even more contentious question of the extent to which
animals may have subjective feelings presents scientists with
considerable problems. For example, intense fear, produced by
exposure to novel situations may produce effects that persist
for the whole duration of the life of the animal. As Griffin
(1976) points out: "It is very easy for scientists to slip into
the passive assumption that phenomena with which their customary
methods cannot deal effectively are unimportant or even non-
existent".

A promising method of investigation is to set up situations
in the laboratory in which the animal is able to select and
choose its own environment. For example, operant methods have
been used to determine illumination preference in pigs and
calves (Baldwin and Start, 1981). It is proposed to extend this
work by exposing animals to simulated transportation and allowing
them to choose by (a) behavioural and (b) operant methods, the
amount of the vibration and noise effects of the simulator, to
which they are exposed.

CONCLUDING REMARKS

Transportation is normally a novel experience for the
animal and this results in behavioural and neuroendocrine
changes. It appears therefore that although transportation is
carried out only occasionally during the relatively short life
of domestic animals, the effects can often be profound. The
degree of response induced in the animal also appears to be
influenced by the nature and duration of each phase of the
journey.

A perplexing problem that confronts clinicians is to
identify which components of the environment are responsible
for the development of clinical signs shortly after completing
a journey. It is also difficult to explain why transportation
leads to problems in some cases while not in others. For

example, in our own laboratory moving sows to be weighed at the beginning of a series of experiments designed to study their metabolism resulted in foetal death on three occasions. Similar subsequent weighings of sows from the same herd produced no detrimental effects. Again, exposure of rabbit does at various stages of gestation to simulated transport produced no untoward effects on the pregnancy. Therefore, the problems associated with experimentally investigating the effects of transportation on pregnancy are complex. It appears that each individual species will need to be considered separately in relation to the effect of transportation on their reproductive performance.

ACKNOWLEDGEMENTS

I wish to thank Dr. A. Glücksmann, of this Institute, for help in preparing the manuscript, Messrs. Stubbs, Ellern Mede, Totteridge Common, London N2O 8LS, for their constructive comments and also Miss C.E. Delaney for skilful technical assistance. Mr. J.M. Stephens, Flower Hill, Tumble, Llanelli and Mr. A.B. Gill, The Abattoir, Coppice Road, Willaston, Nantwich, very kindly allowed unfettered access to the calves for blood sampling at Carmarthen and Nantwich respectively. Mr. B.M. Williams, MRCVS, DVSM, Deputy Regional Veterinary Officer, Ministry of Agriculture, Fisheries and Food, Hook Rise South, Tolworth, Surbiton, Surrey, arranged the taking and estimating of the plasma urea levels in the calves before and after transportation.

Finally, I would like to thank Dr. R.J. Bywater, Ph.D., M.Sc., BVMS, MRCVS, Head of Applied Veterinary Pharmacology, Beecham Pharmaceuticals Research Division, Animal Health Research Centre, Walton Oaks, Dorking Road, Tadworth, Surrey, for supplying the calf milk (Vitamealo), the specially designed bag for collecting faeces samples and for many helpful discussions about the general problems associated with calf rearing, and the aetiology of calf scour in particular.

REFERENCES

Adams, C.E. and Stephens, D.B., 1980. Observations of the effects of simulated transport stress on pregnancy, parturition and respiration in the rabbit. Appl. Anim. Ethol. 6, 390-391.

Agnes, F. and Genchi, C., 1972. (Physiological disorders of cattle after transportation. II Serum proteins, glycaemia, azotaemia, uricaemia). Clinica Veterinaria 95, 3, 65-70.

Averill, R.L.W., 1964. Ovulatory activity in mature romney ewes in New Zealand. NZ J. agric. Res. 7, 514-524.

Bywater, R.J., 1977. Evaluation of an oral glucose-glycine-electrolyte formulation and amoxicillin for treatment of diarrhoea in calves. Am. J. Vet. Res. 38, 12, 1983-1987.

Bywater, R.J., 1980. Comparison between milk deprivation and oral rehydration with a glucose-glycine-electrolyte formulation in diarrhoeic and transported calves. Vet. Rec. 107, 549-551.

Bywater, R.J. and Woode, G.N., 1980. Oral fluid replacement by a glucose glycine electrolyte formulation in *E coli* and rotavirus diarrhoea in pigs. Vet. Rec. 106, 75-78.

Cannon, W.B., 1929. 'Bodily changes in pain, hunger, fear and rage: an account of recent researchers into the function of emotional excitement'. 2nd Edition. Appleton, New York.

Close, W.H., Stephens, D.B., Polge, E.J.C. and Start, I.B., 1981. The effects of transportation, the proximity of a boar and other stimuli on the attainment of puberty in gilts. Appl. Anim. Ethol. In press.

Courrier, R. and Cologne, A., 1951. Cortisone et gestation chez la lapine. C.R. Lebd. Seanc. Acad. Sci. Paris 232, 1164-1166.

Dauncey, M.J. and James, W.P.T., 1979. Assessment of the heart-rate method for determining energy expenditure in man, using a whole-body calorimeter. Br. J. Nutr. 42, 1-13.

DiSalvo, J. and Fell, C., 1971. Changes in renal blood flow during renal nerve stimulation. Proc. soc. exptl. Biol. Med. 136, 150-153.

Doney, J.M., Smith, W.F and Gunn, R.G., 1976. Effects of post-mating environmental stress or administration of ACTH on early embryonic loss in sheep. J. agric. Sci. 87, 133-136.

Feigl, E., Johansson, B. and Löfving, B., 1964. Renal vasoconstriction and the 'Defense Reaction'. Acta physiol. scand. 62, 429-435.

Gray, J.A., 1971. The psychology of fear and stress. Weidenfeld and Nicolson.

Holt, A.F., 1962. Movement of cattle and its effect on fertility. Br. vet. J. 118, 293-298.

Howarth, B.Jr. and Hawk, H.W., 1968. Effect of hydrocortisone on embryonic survival in sheep. J. Anim. Sci. 27, 117-121.

Johnston, J.D. and Buckland, R.B., 1976. Response of male holstein calves from seven sires to four management stresses as measured by plasma corticoid levels. Can. J. Anim. Sci. 56, 727-732.

Lee, J. and Ring, P.A., 1956. The effect of maternally administered cortisone and ACTH upon the pancreas of the foetus. J. Endocrin. 14, 284-291.

Mancia, G., Baccelli, G. and Zanchetti, A., 1974. Regulation of renal circulation during behavioural changes in the cat. Am. J. Physiol. 227, 3, 536-542.

Mandl, A.M. and Zuckerman, S., 1952. Factors influencing the onset of puberty in albino rats. J. Endocrinol. 8, 357-364.

Moberg, G.P., 1976. Effects of environment and management stress on reproduction in the dairy cow. J. Dairy Sci. 59, 1618-1624.

Pappenheimer, J.R., 1960. Central control of renal circulation. Physiol. Rev. 40, Suppl. 5, part II, 35-37.

Rader, R.D. and Stevens, C.M., 1974. Renal parameters estimated in unrestrained dogs. Medical and Biological Engineering, 465-478.

Reid, R.L. and Mills, S.C., 1962. Studies on the carbohydrate metabolism of sheep. Aust. J. agric. Res. 13, 282-295.

Robson, J.M. and Sharaf, A.A., 1952. Effect of adrenocroticotrophic hormone (ACTH) and cortisone on pregnancy. J. Physiol. 116, 236-243.

Selye, H., 1950. The physiological and pathology of exposure to stress. Montreal.

Shaw, K.E. and Nichols, R.E., 1964. Plasma 17-Hydroxycorticosteroids in calves - the effects of shipping. Am. J. Vet. Res. 25, 252-253.

Signoret, J.P., 1970. Sexual behaviour patterns in female domestic pigs (Sus scrofa L.) reared in isolation from males. Anim. Behav. 18, 165-168.

Staples, G.E. and Haugse, C.N., 1974. Losses in young calves after transportation. Br. Vet. J. 130, 374-379.

Stephens, D.B. and Rader, R.D., 1981. The effects of simulated transport and handling on heart rate, blood pressure and renal arterial blood flow in the pig. Appl. Anim. Ethol. In press.

Stephens, D.B. and Toner, J.N., 1974. A method for the continuous display of heart rate in the calf during transportation. J. Physiol. 242, 24-25P.

Stephens, D.B. and Toner, J.N., 1975. Husbandry influences on some physiological parameters of emotional responses in calves. Appl. Anim. Ethol. 1, 233-243.

Thurley, D.C. and McNatty, K.P., 1973. Factors affecting peripheral cortisol levels in unrestricted ewes. Acta Endocrinol. 74, 331-337.

Toner, J.N. and Stephens, D.B., 1975. Instantaneous heart rate indicator with digital display. Laboratory Practice, July 1975, 464.

DISCUSSION

J.P. Signoret *(France)* When you reported your results concerning the accelerated puberty in gilts you mentioned contact with mature stock. Did this involve adult females only, or were there males as well?

D.B. Stephens *(UK)* Both. I should have pointed out that there was no actual physical contact. They were placed in adjacent pens.

J.P. Signoret Males were present, then, not very far away? There was not only contact with adult females. You had no groups of gilts which were about to enter puberty which had contact only with mature, cycling sows?

D.B. Stephens No. You think that would be worth doing? We are trying to do the experiment again, with younger gilts. Do you think that would be important?

J.P. Signoret I do not know. The influence of boars and transport has already been reported as inducing puberty in gilts. Work done on the wild boar, especially intensive work done in the DDR, has shown that within a social group of sows there is a synchronisation of oestrus and this necessitates some kind of inter-female stimulation and not only boar stimulation.

A. Hoogerbrugge *(The Netherlands)* I think this has also been done in the UK. It was shown that there was an influence of the boar. There was a difference between a young boar and an older boar of ten months old. There is also Australian work on the induction of oestrus and on the mating rate. It was shown that there was a significant difference in mating rate when gilts were in contact with the boar from ten weeks of age. So it is not just a question of coming in heat, there is also the willingness of the gilts to be mated. There was a significant difference in this depending on whether they had grown up in contact with the boar or not. So I agree that a lot of stimuli might be involved in this question of coming on heat.

RESEARCH ON TRANSPORT STRESS AT ARC BRISTOL

D. Lister

ARC Meat Research Institute,
Langford, Bristol BS18 7DY, UK.

ABSTRACT

It is no longer sufficient to examine aspects of the welfare of farm animals simply in subjective ways. But the recognition of suitable, objective criteria is difficult. At one extreme, animals may die during transportation, but to many, the obvious consequence of transport stress may be a slight reduction in offal or carcase yield or in the commercial quality of the meat. Few problems are predictable in their outcome and most are recognisable only after the event. It is, therefore, relatively easy to identify conditions which have been stressful but not to predict them. Apart from the conditions imposed on them during handling, animals differ according to species, breed and sex in their responses. This paper, therefore, reviews research on the mechanisms which allow animals to respond to the stresses they are likely to encounter rather than defining the optimum conditions for the transportation of animals.

INTRODUCTION

Of all the farm species, pigs present the greatest risk of sudden death and of these fatalities 70 - 80% occur during transportation. Ludvigsen (1957) described the symptoms of the condition in pigs. Dyspnoea, cyanosis and hyperthermia are common findings and result predominantly from anaerobically supported muscle metabolism. The high concentration of lactate and carbon dioxide liberated into the plasma stimulates the secretion of catecholamines and further anaerobiosis. *Post mortem* examinations usually reveal symptoms of cardiac failure in otherwise healthy animals. The same picture is found, though less commonly, in other farm animals. The first conclusion that we may draw about transport stress is, therefore, that the consequences are more striking in some species than others. Further enquiry reveals that among pigs, and there is anecdotal support in sheep and cattle also, different breeds show different levels of sensitivity and degrees of response. In consequence, given the same transport stress, some animals will die whilst others survive without obvious detriment. A third group, however, may survive but defects such as PSE (pale, soft, exudative/watery) or DFD (dark, firm and dry/dark cutting) meat are obvious in the carcase.

Transport stress, leading to unexpected death, may, therefore, be considered to be a contributor to, or perhaps a final, fatal trigger of what is described as the stress syndrome. Metabolically there seems to be little to differentiate it from reactions to exhaustive physical exercise, for example, emotional or even heat stress.

The common physiological and metabolic components of the general syndrome are often interpreted as the consequences of profound stimulation leading to metabolic and respiratory acidosis through the progressively increasing contribution of anaerobic glycolysis. There are additional consequences on muscle which result from the provision of specific energy substrates and their 'preferred' utilisation in metabolism.

Work at the ARC Meat Research Institute has been conducted along three routes, the first of which has been to examine the common, basic mechanisms involved in the initiation and propagation of metabolic disturbances in animals. Secondly, we have researched their relevance in practical and commercial situations which are considered to be stressful. Thirdly, we have attempted to mimic in experiments the consequences of certain responses to stress and to counteract them with various drugs.

PHYSIOLOGICAL MECHANISMS

The telemetric monitoring of some physiological information from unrestrained animals reacting to stressful conditions is now possible. For example, heart and respiration rates, muscle movements, temperature and pH of some body fluids can be measured. It is also possible to implant cannulae in arteries and veins to allow blood sampling. Even so there are many problems to be overcome before these techniques can be used satisfactorily on animals progressing to slaughter. There has, therefore, been some need to identify a model by means of which responses to stress can be examined experimentally.

The reaction known as the Malignant Hyperthermia (MH) syndrome has provided such a model for investigations of the initiation and propagation of gross metabolic disturbances. The syndrome, which may develop spontaneously, notably in pigs but also in some other species, including man, as a response to

physical or emotional stimulation can be induced almost at will
with the use of certain anaesthetic agents. MH has been
described in many publications, most recently by Gronert (1980).
Its relevance in the present context is supported by the following in particular:

a)   the common involvement of the sympatho-adrenal system;
b)   the progressive contribution of anaerobiosis in MH and
      stress reactions;
c)   the specific roles of liver, muscle and adipose tissue
      in supporting energy metabolism as the reactions proceed.

The physiological mechanisms involved in all these give
rise to symptoms which can be and are used in attempts to
characterise in objective terms the degree of stress imposed in
a given environment. Thus the first notable symptom is an
increase in heart rate. Respiratory and metabolic acidoses are
associated with changes in rates of respiration. There may be
an increase in body temperature and there are changes in plasma
electrolytes, notably potassium, which eventually contribute to
the death of the subject.

The importance of the rises in circulating catecholamines
in propagating if not initiating MH is shown in the effects of
adrenalectomy and sympathetic blockade. Animals so treated
will not respond to the usual triggering drugs (Lucke et al.,
1978) and prior treatment with the α-adrenergic blocking drug,
phentolamine (Lister et al., 1976) will prevent the development
of an MH reaction. Anti-serotonin drugs will apparently do the
same (L. Ooms - personal communication 1981). It is also known
that tranquillisers and neuromuscular blockade will attenuate
and even prevent a response in some susceptible animals, but
not all. Conversely a few minutes mild exercise may increase
the apparent proportion of reactors in a population (Van den
Hende et al., 1976). The only known drug to correct an established MH reaction and affect a cure is dantrolene sodium
(Gronert, 1980).

Hepatic metabolism in MH has been extensively investigated
(Hall et al., 1980). There is no evidence of abnormal hepatic
function and so long as blood flow can be appropriately maintained glycogenolysis can maintain appreciable glucose efflux

from the liver (and give rise to hyperkalaemia). There is a
decreasing contribution of free fatty acids (through re-esterification) to the support of metabolism which eventually must
create a significant drain on the alternative energy substrates
of liver and muscle glycogen and cause their eventual dis-
appearance from the tissues.

With this general picture of events in MH it is possible
to predict and evaluate changes likely to occur in animals subjected to stresses such as those they might encounter during
transportation.

THE ASSESSMENT OF REACTIONS IN PRACTICAL SITUATIONS

The investigation of MH has shown how important the
sympatho-adrenal system is in stress or drug induced metabolic
reactions. If there were one separate mechanism for each type
of stress response it would be relatively easy to identify the
physiological components of, say, transport stress. Unfortunately, it is difficult to separate the effects of mixing, loading, transporting, unloading, the inevitable period of fasting
which accompanies transportation and any recovery which may
occur in lairing - all of which involve to a greater or lesser,
and sometimes conflicting, extent, the actions of catecholamines.
Thus the mild hypoglycaemia to be expected from fasting may be
offset by a hyperglycaemic response to physical stress. Low
concentrations of free fatty acids (FFA) in the plasma resulting
from their re-esterification may obscure the consequences of
active lipolysis, as measured by plasma glycerol concentrations,
if the subject is even mildly acidotic. These kinds of compli-
cations have been observed in blood taken from groups of commercial cattle at slaughter (Warriss, 1981 - in preparation).

The only significant changes in blood metabolites were
small and included increased concentrations of ketones, and
decreased levels of lactate and amino acids in the serum. Overall the results suggested that the cattle tended towards a state
of negative energy balance, but the main conclusion was that
analyses of samples of blood taken at slaughter are difficult to
interpret and may be quite misleading.

Serial sampling of blood taken via chonically implanted cannulae may be more useful in following the responses of animals subjected to the stress of transportation and associated handling In his experiments, in which he followed the effects of a cycle of events - loading, transportation and unloading - on various hormones and metabolites, Spencer (in press) concluded, contrary to several other observers, that transport provided additional stress over and above that due to loading and unloading.

All this goes to show not that one author is right and another wrong but how difficult it is to identify specific physiological mechanisms and associations in populations of animals under p. actical conditions. Not only is the genotypic variation of the population under scrutiny likely to be large, but the experiences, behaviour and physiological states of individual animals from a single group may be very different. Thus we might be able easily to measure by such surveys the consequences of treatment but not the means whereby it was brought about. An example of this comes from recent other work in our laboratory (Warriss, 1981 - submitted) which clearly shows the combined effects of fasting and transport on the losses of body tissue weights of pigs. Carcase and liver weights are reduced in proportion to the length of fast after the post absorptive phase of feeding begins (16 h). Muscle and liver glycogen depletion is a major contribution to weight loss but also, in the case of muscle, to the raising of ultimate pH. We have also observed this effect as a difference between the effects of transportation on Friesian bulls and steers.

THE EXPERIMENTAL INDUCTION AND PREVENTION OF STRESS REACTIONS

The relative unprofitability of the survey approach in identifying mechanisms responsible for the responses of animals to pre-slaughter handling has led us to lay greater emphasis on work which attempts to mimic or correct the commonly observed responses to stress or their purported causes. Because of the pertinence of responses which involve the sympatho-adrenal system we, naturally, have given much attention to this topic.

We (Gregory and Lister, 1981) have shown that the greater sensitivity to stress of breeds of pig like the Pietrain is

associated with the release of greater amounts of catecholamines
for a given stimulus than is seen in stress resistant pigs such
as Gloucester Old Spots. But it seems that the cardiovascular
system of Pietrain pigs is not unduly sensitive to the effects
of catecholamines.

Experience in the control of MH suggests that α-adrenergic
blockade prior to exposure to the triggering drug will prevent
a reaction from developing. It will not control an established
one. β blockade will do neither (Lister et al., 1976). It is,
however, the β-adrenergic actions on lactate production which
are of such consequence in stress reactions and a purported
β-blocking drug (carazolol; Suacron, Praemix/Wirkstoff GmbH,
Mannheim) has been used most effectively in preventing PSE meat
from developing in transported pigs (Warriss and Lister - in
press). Of relevance here is the potency of carazolol which is
approximately one hundred times greater than that of the pro-
pranolol used in the MH experiments (Lister et al., 1976).

β-adrenergic stimulation with iso-prenaline infusions is
sufficient to induce acidosis, muscle glycogen depletion and
dark cutting meat in stress prone pigs, but in ruminants it may
be necessary to reduce FFA even further with the parenteral
administration of nicotinic acid or methyl pyrazole carboxylic
acid (MPCA) (Lister and Spencer, 1981 - in preparation). High
pH meat may develop by these means even though there may be
appreciable quantities of glycogen remaining in the liver. On
the other hand, α-adrenergic stimulation with intravenous
phenylephrine has marked effects on liver glycogen in pigs and
ruminants, but only after prolonged infusions, when glycogen
has almost disappeared from the liver, is there any evidence of
muscle glycogen depletion.

These results taken together suggest that reactions to
stress are the consequence of both α- and β-controlled adrenergic
and adrenomedullary mechanisms. Terminally the β-anaerobiosis
stimulating component appears to predominate.

Although its exact role with respect to hormone secretion
by the adrenal gland is not known, the amounts of ascorbic acid
present in the adrenal have been shown to reflect the stress

suffered by an animal prior to slaughter and the extent of
recovery possible through lairing (Warriss, 1979).

Dantrolene sodium, a hydantoin derivative, is the only .
consistently effective agent as prophylaxis and in the treatment
of MH. It is also effective in the prevention of PSE in pigs
which have been mixed and transported prior to slaughter (Lister
and Spencer, 1981 -in preparation). It is thought to exert its
action on muscle via calcium transport mechanisms. At first
sight, therefore, it is not obviously involved in the mechanisms
implicated above. However, whereas β-adrenergic stimulation
always occurs via the adenyl cyclase - cyclic AMP system, α-
stimulation depends on changes in the intracellular calcium
concentration (Thompson and Williamson, 1976).

FURTHER WORK

It is difficult to imagine a code of practice for the
handling of animals prior to slaughter which can eliminate all
risk of their encountering situations which could lead to
untimely deaths or meat of reduced quality. The nature of the
spectrum of sensitivity of stress amongst animals ensures that
extreme steps taken to protect them, including the use of potent
drugs, will not guarantee universal success. All the care taken
during the pre-slaughter handling can be reduced to nought by a
momentary lapse of the operative's attention at a critical point
during slaughter. Finally, we can only have indirect evidence
that a particular practice is both foolproof and humane.

This is not to say that we should stop searching for
better ways of handling animals. But we should ensure that the
animals we raise can withstand the handling systems or environments they might reasonably be expected to endure or cannot
easily be protected from.

A first step to reduce the stress associated with transportation, which is supported strongly by economic arguments
alone, would be to see that animals spend the shortest practicable time in transportation and pre-slaughter handling.

So far as physiological investigations are concerned,
there is an urgent need to discover the nature of the differences in sensitivity between animals. We have already made some

progress in this with pigs. The responsiveness of the pig's sympathetic nervous system can be tested, as it can in man, with a valsalva-like-manoeuvre (Gregory and Lister, 1981). In ruminants, gut fill leads to difficulties (Gregory et al., 1981) and other alternative approaches are clearly needed. The identification of sensitivity thresholds is of appreciable importance for it will allow us to use physiologically comparable animals in the assessment of environmental stress. It will further allow us to attempt to differentiate and perhaps separate the purported linking between sensitivity to stress and characteristics of the growth of animals (Lister, 1980).

The central part played by anaerobiosis in precipitating premature death or meat quality problems clearly pinpoints the need for research into how it can be blocked or delayed in onset in stressful circumstances. The relative ease with which anaerobic muscle metabolism is enlisted may depend primarily on the characteristics of the autonomic nervous system and it is a function of the successful perfusion of tissues with oxygenated blood. There are many ways of maintaining adequate peripheral blood flow with the use, for example, of potent adrenergic blocking drugs. But in the longer term we shall need other solutions which leave no problems of residues in their wake.

It has been shown how important is the time between the last meal and slaughter in modifying carcase, tissue, and organ weights. Most of these changes represent direct or indirect losses of nutrients from the body. There have been efforts made to replenish these losses by feeding sugar or molasses prior to slaughter, but there has been little interest in the feeding of specific nutrients to animals prior to despatch from the farm as a possible means of protecting them from the consequences of stressful encounters. The success of treatments designed to raise glycogen stores in endurance athletes is well documented as an example of a benefit of this approach.

It is clearly evident, however, that the elimination of stress problems will not be accomplished simply by better understanding of the physiological principles involved. Such knowledge is, of course, paramount, but remedies and solutions will be identified by a combination of many approaches perhaps

involving behaviourists and engineers as well as physiologists
and meat scientists.

REFERENCES

Gregory, N.G., Audsley, A.R.S. and Lister, D., 1981. Studies on the sympa-
thetic nervous system: the Valsalva like manoeuvre in sheep. Res.
Vet. Sci. 30, 284.
Gregory, N.G. and Lister, D., 1981. Sympathetic responsiveness in relation
to fatness in pigs. Proc. Nutr. Soc. 40, 11A.
Gronert, G.A., 1980. Malignant Hyperthermia. Anesthiology 53, 35.
Hall, G.M., Lucke, J.N., Lovell, R.D.L. and Lister, D., 1980. Porcine mal-
ignant hyperthermia VII. Hepatic metabolism. Br. J. Anaesth. 52, 11.
Lister, D., 1980. Hormones, metabolism and growth. Reprod. Nutr. Develop.
20(1B), 225.
Lister, D., Hall, G.M. and Lucke, J.N., 1976. Porcine malignant hyperthermia
III. Adrenergic blockade. Br. J. Anaesth. 48, 831.
Lucke, J.N., Denny, H.R., Hall, G.M., Lovell, R.D.L. and Lister, D., 1978.
Porcine malignant hyperthermia VI. The effects of bilateral adrenal-
ectomy and pre treatment with bretylium on the halothane-induced
response. Br. J. Anaesth. 50, 241.
Ludvigsen, J., 1957. On the hormonal regulation of vasomotor reactions
during exercise with special reference to the action of adrenal cort-
ical steroids. Acta. Endocr. 26, 406.
Thompson, M.R. and Williamson, 1976. Metabolic effects of alpha- and beta-
adrenergic stimulation of rat submaxillary gland *in vitro*. Biochem.
J. 160, 597.
Van den Hende, C., Lister, D., Muylle, E., Ooms, L. and Oyaert, W., 1976.
Malignant hyperthermia in Belgian Landrace pigs rested or exercised
before exposure to halothane. Br. J. Anaesth. 48, 821.
Warriss, P.D., 1979. Adrenal ascorbic acid as an index of pre-slaughter
stress in pigs. Meat Sci. 3, 281.

DISCUSSION

R. Dantzer *(France)* I wonder whether we need a model, at the moment,
concerning transport stress in which we are able to elucidate the key mechanism inducing transport stress? In other words, I think that at present
we are confronted with important practical problems which can be solved by
recourse to such a model.

My second comment concerns the validity of the malignant hyperthermia
syndrome as a model for stress in pigs. I am deeply convinced that stress
is not just the animal's physiological response to something in the environment, whatever that thing in the environment may be, chemical or physical.
I think that stress is the result of cognitive processes and hormonal reactions. It is an interplay between these two things. I agree that you can
study physiological mechanisms, but you cannot find such mechanisms which
do not enable the animal to withstand some kind of stimulation. The way the
animal will react does not depend only on the physiological mechanism *per se*.
We have done experiments on Piétrain positive pigs, Piétrain negative pigs
and Large White pigs, and in our view they are no more sensitive to stress
than you and I. Stress is not only the result of physiological mechanisms,
but it is also the result of behavioural mechanisms.

You have shown that sympathetic nervous system activity differs in malignant hyperthermia sensitive pigs, but the results you have shown are results which compare Piétrain pigs either to Large White or to Gloucester pigs. Do the same results apply to Piétrain positive pigs when they are compared to Piétrain negative pigs?

D. Lister *(UK)* I thought I had made it absolutely clear that there is no way of doing investigations on stress without recourse to a model system, because we have been dealing with such muddy water for the last 40, 50 or even perhaps 200 years by simply looking at animals and saying, 'Oh, yes, that is more stressed than that'. If you are going to prove anything, or come up with concrete questions to identify what stress means, what welfare criteria are, you have got to give numbers in some way. You have to say this is different from that, this treatment is different from that one, because we can identify certain things which are occurring and others which are not occurring. I tried to demonstrate that the outcome of the model could be used as a detective story to enable us to recognise what happens to an animal which is being transported, kicked or whatever. You have to iden-tify how animals respond in very objective ways. I will not accept subjective views, as has been the practice for so very long. We have to have concrete information, and this is why the welfare debate goes on *ad nauseam*. There are not enough objective data.

No, we have not compared halothane positive and halothane negative pigs in these ways. I have no reason to believe that there are sizeable differences between halothane negative and halothane positive animals. I would argue with you very strongly about what constitutes a halothane pos-itive and a halothane negative pig, because if you are just going to tell me that a halothane positive pig is one which reacts in three minutes I would deny it. We have Piétrain pigs and Vincent McLoughlin has halothane positive pigs which take hours to react. I would be very surprised to see a very clearcut definition between halothane positive and halothane negative. Some Piétrains take longer than others. There is a spectrum of reaction. Some Large Whites will react and others will not in the time allotted. By walk-ing Belgian Landrace pigs around or exciting them for a period you can make all of them react, whereas if they were kept quiet they would not react, it seems.

I cannot answer your question directly, because we have no comparison of Piétrain positive and negative, but I would argue very strongly that the spectrum of sensitivity is across all pigs rather than just across Piétrain positive and negative.

To turn to your second point, I think it is up to you to consider whether malignant hyperthermia is a model for stress. The whole point of using models is to demonstrate whether the outcome is relevant to the sit-uation you are modelling. I would argue that it is.

R. Dantzer I can give you another example if you like. It is well known that some people are born with metabolic defects in certain enzyme systems which lead to mental retardation. You cannot say that the particular enzyme is not necessary for mental development, but neither can you say that lack of it is a symptom, model, for mental disease or mental retardation.

D. Lister All you are doing is choosing an inappropriate model. I have tried to choose an appropriate model.

R. Dantzer In other words, in your model you show that the sympathetic nervous system is important, and everybody knows that this is so, but this does not mean that it is a model of stress.

D. Lister How have they demonstrated that the sympathetic nervous system is important? You see, if you take animals in transport and measure catecholamine, you can have any value you like. I would not suggest anything on the basis of results from half a dozen animals transported across the town in which the catecholamine levels were measured, because they are so misleading and difficult to interpret. That is the whole point of identifying a model - to look to see what happens when you stimulate it in a particular way.

W. Sybesma *(The Netherlands)* I should like to ask Dr. Lister whether there is any evidence for a link between the irritation of the sympathetic nervous system and the potential for muscle growth.

D. Lister I would like to think that there was, but there is no evidence that this does occur. It seems strange to me that the animals in which there is a problem always seem to be those which have the ideal body composition. In other words, they always seem to be the least fat animals. That is another story into which I would happily go.

N.J. Nielsen *(Denmark)* I should like to ask Dr. Lister which treatment he would recommend in practice to minimise PSE meat. Should the pigs be fed or not on the day of slaughter? What about resting time - immediately after arrival or after 2 - 4 h - and what about fighting?

D. Lister I could have given you some slides of an experiment we did comparing the effect of removing food up to 48 h before slaughter and coupling that with lairing for 1 h or 24 h before slaughter. There is no doubt that the longer the delay from the last feed to the time of slaughter, the lower the killing out percentage becomes and the loss of glycogen from the liver and the muscle becomes very obvious indeed. In other words, there is a rise in the ultimate pH of meat the longer you keep the animal without food. I think you could never advise a producer to starve his animals for a substantial period of time because he will lose money very rapidly indeed, once 16 h have elapsed. You can normally allow an overnight fast and the animal will not have lost substantial amounts of weight and the ultimate pH of the meat will not have been affected. After that time, the problems become even greater. I would simply say that, so far as pigs are concerned (and I take the point that Mr. Platt made this morning about animals which have been transported for several days), there is no justification for complicating its life with an additional four hours in lairage. I would happily take two hours as the time which is reasonable to have animals awaiting slaughter.

We have to remember that the whole point of keeping animals in lairage is simply to provide a constant flow of animals onto the slaughterline. It was never intended that it should become a recovery period. It was simply a way of ensuring that a slaughterline could operate for 6, 8, 12 h or whatever. I think that as far as pigs are concerned, they ought to be slaughtered in the very minimum time from the place of production to the place of slaughter.

RESEARCH ON FARM ANIMAL TRANSPORT IN FRANCE: A SURVEY

R. Dantzer

Laboratoire de Neurobiologie des Comportements INRA,
Université Bordeaux 11, 146 Rue Léo Saignat 33076 Bordeaux,
Cedex, France.

INTRODUCTION

In 1979 the French Deputy P. Micaux was nominated by the Government to enquire and advise on animal welfare standards in modern farming. In his report which was published last year, many welfare problems related to transport of farm animals were recognised (Table 1). Most of these problems could be solved by the strict application of the existing legislation and a better co-ordination between the different administrative bodies in charge of transport regulations in France, but there still remained a need for more information on vehicle designs, stocking densities, animal handling conditions and other factors relative to the knowledge of physiological and behavioural needs of farm animals (Table 2).

TABLE 1

MAIN WELFARE PROBLEMS RELATED TO TRANSPORT OF FARM ANIMALS (MICAUX, 1980)

Horses	:	Inadequate transport conditions Lack of qualified staff
Calves	:	Inability of battery calves to stir
Pigs	:	Loading and unloading difficulties
Kids	:	Handling (bound legs)
Poultry	:	Crowding in transport cages

For many species, transport includes one or several stays in livestock fairs and transit centres.

When dealing with transport it is convenient to consider separately the different physiological and pathological consequences of this procedure (e.g. mortality, weight loss, changes in meat quality, blood constituent changes, etc.). In France, as in other countries, research has proceeded along these lines and will therefore be presented accordingly. However

TABLE 2

MAIN SOLUTIONS FOR ENSURING THE WELFARE OF TRANSPORTED ANIMALS IN FRANCE
(MICAUX, 1980)

1.	Elevator for pig and calf containers
2.	Hydraulic loading ramps
3.	Temperature and ventilation standards
4.	Density standards
5.	Inside partitions for separation of individuals (horses)
6.	Hygienic rules
7.	Rules for road and rail transport modalities (speed, transit time, breaks, etc.)
8.	Rules for emergency intervention
9.	Practical rules for securing rest periods and providing food and water to animals in transit, according to their physiological needs.

there is a need at the present time for a more integrated
approach taking into account at the same time the different
consequences of transport, their mutual relationships and their
dependence on the transport process.

The aim of the present report is (1) to give examples of
the work which has been done to evaluate the adverse consequences
of transport of livestock and (2) to point out the areas which
have been neglected and would need further investigation.

RESEARCH ON THE EFFECTS OF TRANSPORT

Statistics on mortality during or shortly after transport
are available only for individual abattoirs or groups of abattoirs, so that it is difficult to get a precise estimate of the
number of transport deaths in the different farm animal species
and to know what are the contributing factors. There is a need
for a centralisation of the data collected by abbatoirs,
insurance companies and producer organisations.

Loss in weight during transport has been the subject of
extensive studies both in pigs and in calves. In pigs transported
from nurseries to the fattening units, Dantzer (1970a) found
average losses of 2% in liveweight after a journey of 2 h (about
100 km). Excretion of faeces and urine accounted for 41% of the
total loss. When corrected for excretory losses, weight losses

were shown to be correlated to ambient temperature and humidity
(Figure 1). This confirms that most of the loss is represented
by water lost through the skin and the respiratory tract, the
remainder being related to respiratory exchanges (for a respiratory quotient of 1, 0.5 g is lost for each litre of oxygen
consumed). Comparison of the weight losses of transported
pigs with those observed in animals submitted to handling but
confined in a motionless lorry during the same time showed that
the road travel by itself was responsible for about 36% of the
total loss in weight while 22% was imputable to handling and
confinement (Figure 2). In the same study, weight losses were
shown to increase with the duration of transport and were
greater when pigs from different social origins were mixed
together.

Although it has been claimed that weight loss can be
reduced by injecting tranquillisers, a systematic study of
the effectiveness of different drugs (phenothiazines and
butyrophenones) in transported pigs demonstrated that the
results were highly variable, depending on the drug used and
the climatic conditions (Dantzer, 1970b).

Weight losses in transported calves amount to 3 - 5%.
The length of the recovery period on arrival in the fattening
unit could be substantially reduced by providing water *ad
libitum* to the animals (Chauvet, 1977). Tissue losses represented 1 percentage point from the carcase, while the liver increased
in weight by 13% (Chauvet, 1981).

Meat quality alterations after transport have been studied
mainly in cattle and in pigs, in relation with the dark, firm
and dry condition (DFD) and the pale, soft, exudative pork
(PSE). The present review will not go into details of the
available studies since the incidence of meat quality alteration
cannot be directly equated to poor transport conditions. One
of the studies worth mentioning is a recent work by Gire and
Monin (1979) in which the authors attempted to assess the role
of stress hormones in the biochemical changes leading to DFD
meat in lambs (Gire and Monin, 1979; Monin and Gire, 1979).
For that purpose, they studied the influence of alpha- and
beta-blocking agents and corticoids on the effects of transport

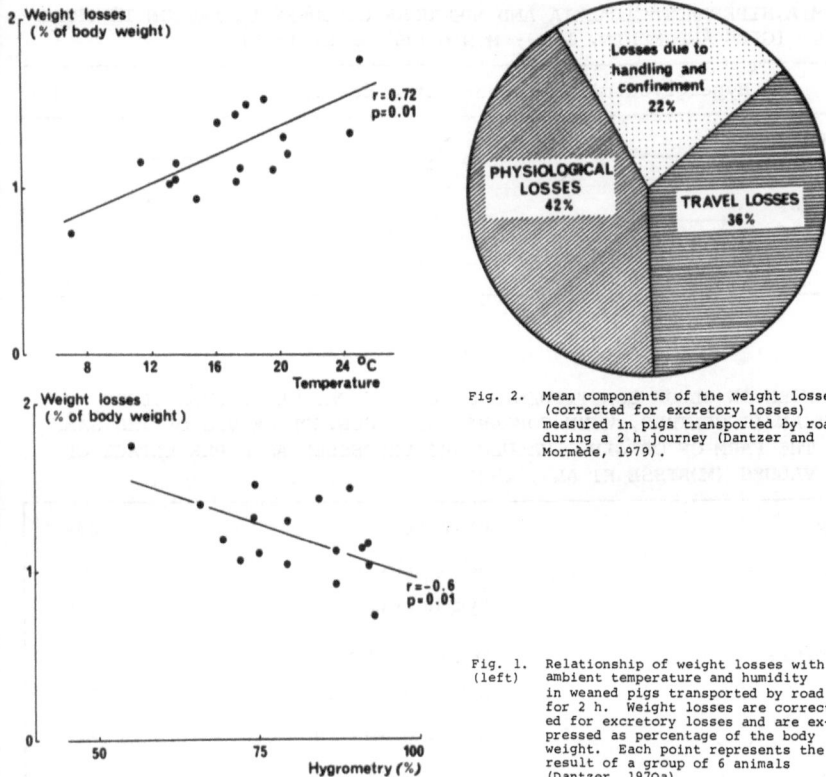

Fig. 2. Mean components of the weight losses (corrected for excretory losses) measured in pigs transported by road during a 2 h journey (Dantzer and Mormède, 1979).

Fig. 1. Relationship of weight losses with (left) ambient temperature and humidity in weaned pigs transported by road for 2 h. Weight losses are corrected for excretory losses and are expressed as percentage of the body weight. Each point represents the result of a group of 6 animals (Dantzer, 1970a).

upon meat quality and compared adrenaline infusion with transport. The results indicate that catecholamines and cort-icoids contribute actively to glycogen mobilisation observed during transport (Table 3). However, the existence of differences between individual muscles in the amount of biochemical alterations when animals perfused with adrenaline were compared to transported animals indicates that physical exercise and/or local blood flow changes play pre-eminent role.

Changes in the physiological status of the transported animals have been mainly studied by measuring blood composition and plasma hormonal levels. Extensive studies are currently underway in calves in order to find physiological criteria which could be used to assess physical exhaustion and suffering in transported animals.

Table 4 summarises the main differences in haematology and blood constituents found in physically exhausted 8 - 15 day old calves, sampled in transit centres after having gone

TABLE 3

SUMMARY OF THE EFFECTS OF ADRENERGIC BLOCKING AGENTS AND CORTICOIDS ON
HYPERGLYCAEMIA, HYPERLACTACIDAEMIA AND MUSCULAR GLYCOGEN DEPLETION IN TRANS-
PORTED LAMBS (GIRE AND MONIN, 1979; MONIN AND GIRE, 1979)

	Hyperglycaemia	Hyperlactacidaemia	Glycogen depletion
Alpha-blockers (hydergine)	←→	↘	←→
Beta-blockers (propranolol)	←→	↘	↘
Corticoids	↗	↗	↗

TABLE 4

MAIN DIFFERENCES IN BLOOD CONSTITUENTS OF 8 - 15 DAY OLD CALVES BLOOD
SAMPLED IN TRANSIT CENTRES, WHEN COMPARED WITH CONTROL CALVES OF THE SAME
AGE KEPT IN THE FARM OF ORIGIN. VALUES ARE EXPRESSED AS A PERCENTAGE OF
THE CONTROL VALUES (MORISSE ET AL., 1981)

Haematology:	Platelet	123%
	Neutrophiles	150%
	Lymphocytes	66%
Serum proteins:	Total protein	91%
	Albumin	93%
	Gamma globulin	58%
Biochemical parameters:	Glucose	68%
	Urea	194%
	Serotonin	157%
	Creatinin	455%

through a market, when compared to calves of the same age kept
in the farm of origin (Morisse et al., 1981). The influence
of the transport itself is apparent from the data shown in
Figures 3 and 4 which represent the changes in blood constituents
observed at different sampling times before, during and after a
12 h transport, with a stop at a transit centre (Mormède et al.,
1981). Transport induced increases in blood urea, creatinine,
bilirubine, SGOT and LDH, suggestive of damages to kidney and
liver. Energetic metabolites (glucose, triglycerides) were
observed to decrease. Some degree of cellular damage was
evident from the increased levels of beta-glucuronidase, a
lysosomal enzyme. For most variables, the changes observed at

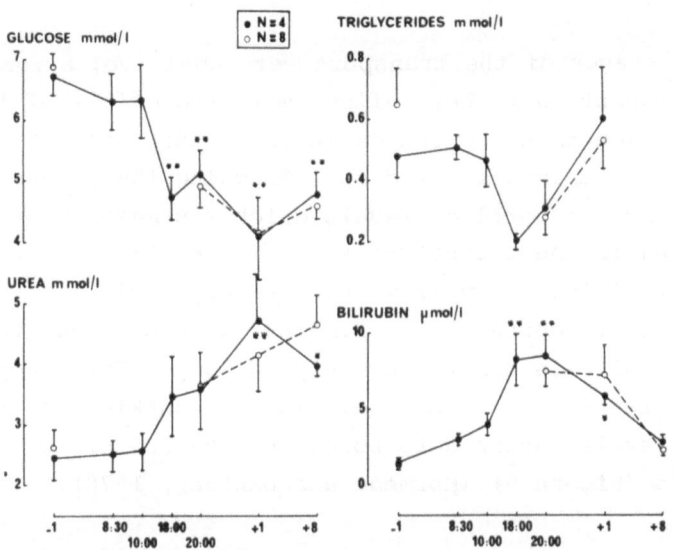

Fig. 3. Main changes in blood constituents of 8 - 15 day old calves submitted to a transport by road from 08.30 to 20.00. Abscissa represent sampling time: -1 refers to blood sampling in the farm of origin (between 14.00 and 17.00) one day before the transport. 08.30, 10.00, 18.00 and 20.00 refer to the time on the day of transport: loading took place at 08.30, the animals were sampled 1½ h later during a stop and then at 18.00, at the arrival in the transit centre. The last blood sample was taken at 20.00 on arrival in the fattening unit. +1 and +8 refer respectively to blood samples taken one day and 8 days later, in the afternoon. Full circles represent the results obtained in 4 sampled calves, open circles the results obtained in 8 calves. (Mormède et al., 1981).

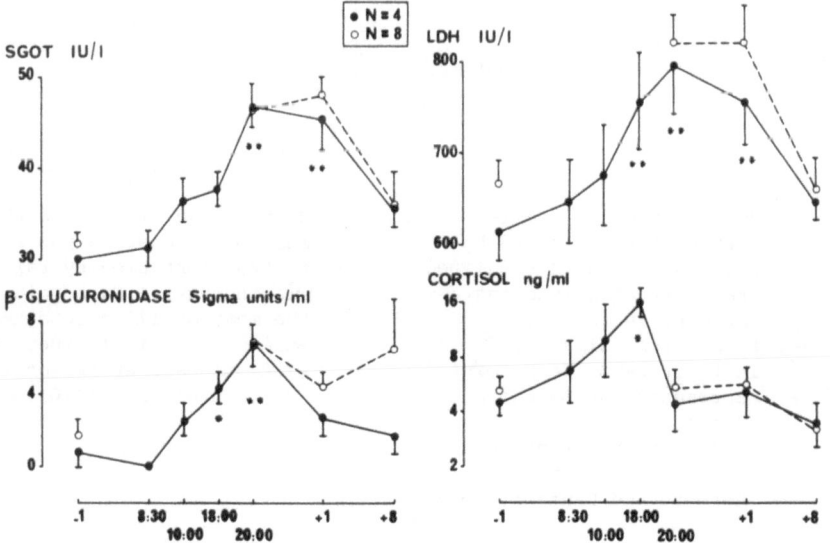

Fig. 4. Mean changes in blood constituents of 8 - 15 day old calves submitted to transportation by road. Same conventions as in Figure 3.

the different stages of the transport were additive, a result which would suggest that they reflect more the effect of the transport duration than the consequences of the different components of the transport process. Trucking the calves produced an increase in cortisol levels which was maximum at the time of arrival in the transit centre. The cortisol levels returned to the initial levels within one day. Plasma cortisol levels are certainly of limited interest since the pituitary-adrenal axis habituates very rapidly. For example in pigs submitted to a transport simulation (shaking for 6 h), the cortisol levels returned to normal by the end of the shaking period (Figure 5) (Mormède and Dantzer, 1978).

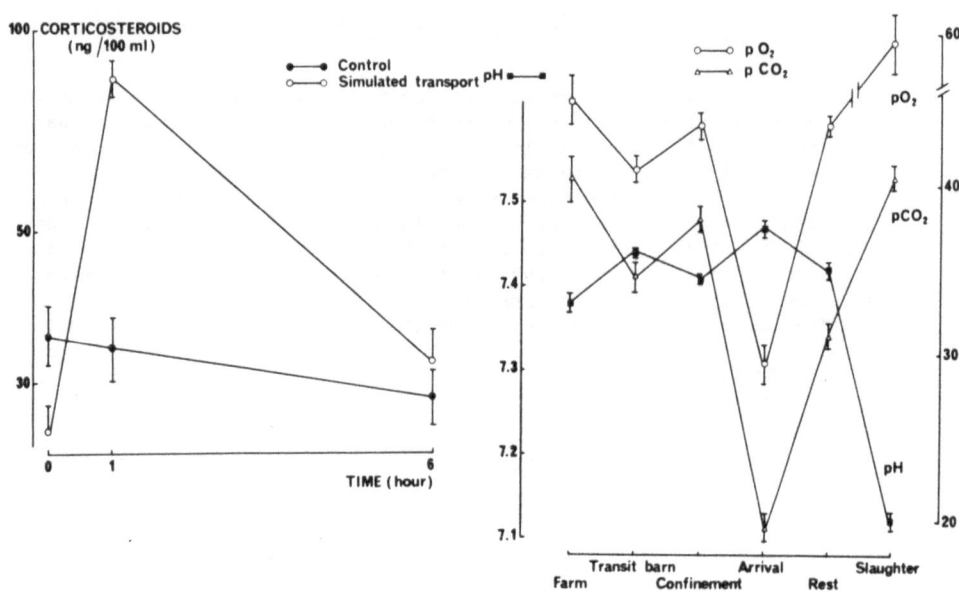

Fig. 5. Mean changes in plasma corticosteroids of 20 kg pigs submitted to simulated transportation (shaking for 6 h) in comparison to control animals left in their pen. Each point represents the mean of 24 pigs. Vertical lines represent the standard error of the mean (Mormède and Dantzer, 1978).

Fig. 6. Mean changes in venous pH, pCO_2 and pO_2 in beef calves submitted to transportation by rail (500 km, one day). Each point is the mean of 17 to 26 determinations. Vertical lines represent standard errors of the mean (Mouthon et al., 1976).

Marked alterations of the acid-base status have been observed in 18 - 30 month old beef calves submitted to a 500 km journey by rail (Mouthon et al., 1976) (Figure 6). Blood pH increased while carbon dioxide tension (pCO_2) and oxygen tension (pO_2) in the venous blood decreased significantly, the greater changes occurring in conjunction with handling (arrival in the transit barn and at the abattoir). These changes are certainly related to a metabolic acidosis partially compensated for by a respiratory alkalosis. However they cannot be used for routine investigation, since they need to be determined immediately, in fresh blood samples.

Reactions to transport stress can also be assessed by reference to the incidence and severity of zootechnical and pathological troubles affecting the animals during the first 2 - 3 weeks of adaptation to their new husbandry conditions. Statistics on mortality and morbidity are available in many producer organisations, but their relationships with transport conditions have been investigated on only a few occasions. For example a close association has been demonstrated between the incidence and severity of abomasal ulcers in calves and both the duration of transport and the travel conditions (Figure 7) (Chauvet, 1977). However correlations are interesting only as far as the true causal factors are determined. Investigations aiming at elucidating the role of transport stress in the pathogeny of transport associated diseases (parasitic outbreaks, transport tetany, shipping fever) have therefore been undertaken. Figure 8 illustrates the case of transport tetany for which the increased sequestration of magnesium and calcium by adipocyte membranes under the influence of catecholamines would be the intermediate factor precipitating the pathological condition (Rayssiguier, 1977).

Reproductive functions can be affected by transportation. Moving of gilts at the age of sexual maturity resulted in appearance of oestrus within a few days (Dumesnil du Buisson and Signoret, 1962). Studies are under way to asses the role of pituitary-adrenal hormones in the appearance of oestrus.

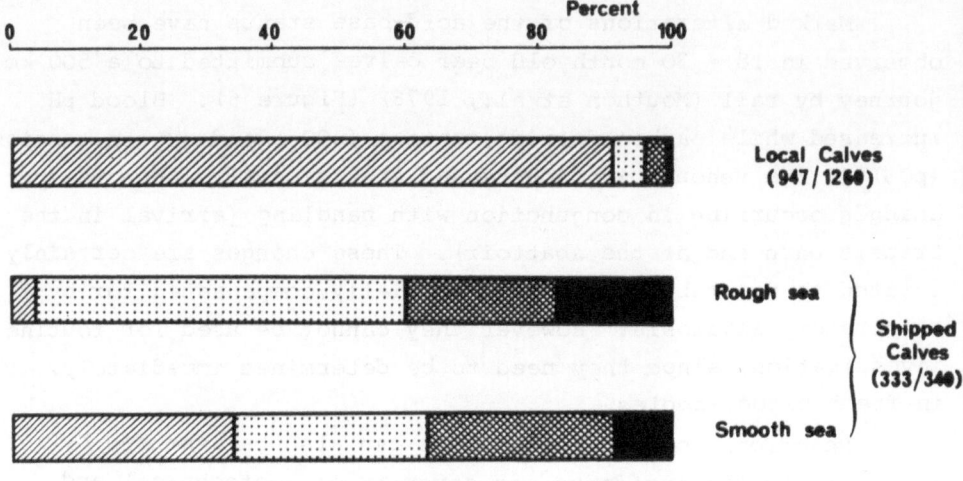

Fig. 7.   Incidence and severity of abomasal ulcers in fattened calves, according to transport conditions. The number in brackets refers to the number of calves showing abomasal ulcers on the total number of calves studied (Chauvet, 1977).

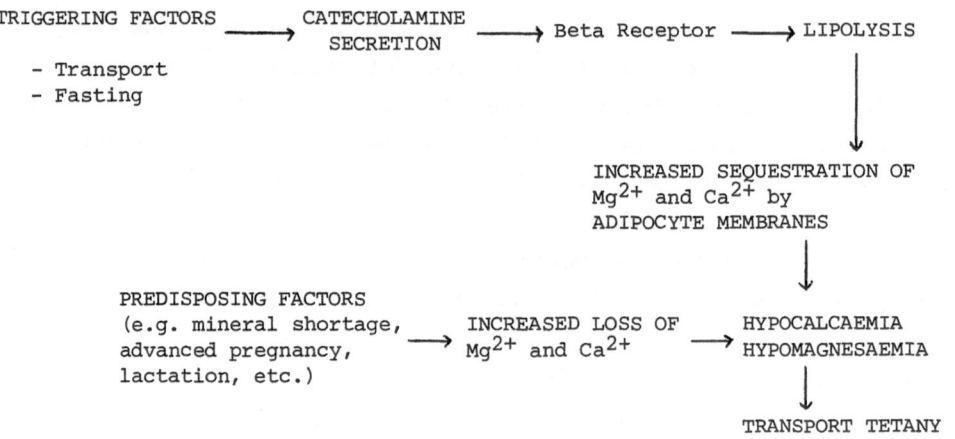

Fig. 8.   Pathophysiology of transport tetany in calves and sheep.

CONCLUDING REMARKS

This review has shown that much research effort has been devoted to assess and analyse the adverse effects of transport in farm animals. However, the results obtained do not necessarily provide all the answers to the practical problems with which welfarists, veterinarians and farmers are confronted. Several reasons account for this discrepancy:

1.   Many problems arise from inadequate transport conditions (e.g. poorly designed vehicles, lack of ventilation or air conditioning, lack of qualified staff to attend the animals, and diluted responsibilities of people involved in transport processes). For example a recent accident which was the source of much suffering was due to the crumbling of the upper floor in a two-deck truck transporting beef calves and waiting at Nice for clearance. The accident occurred at night and the animals were not helped before the late afternoon of the next day (source: AFP and French OABA).

2.   Little research has been devoted to the design of vehicles carrying the animals and the devices which can be used for collecting, loading and unloading. A noteworthy exception is the collection of chickens and guinea-fowls which has given rise to the design of suitable devices both facilitating the loading process and reducing the amount of bruising and injury (Fort, 1979).

3.   Data are lacking on species for which welfare problems may be acute and have no equivalent in other European countries. This is specifically the case for horses which are transported for slaughter over very long distances (Figure 9), but also for other productions such as rabbit meat (Table 5). Another problem is the exportation of live animals. Beef calves exported from France to Italy amount to 700 000 - 800 000 animals/year. This trade has been accompanied by important problems of adaptation on arrival and many losses during and after transit. Preconditioning programmes have been introduced but each marketer has his own solution, the effectiveness of which is very difficult to appreciate. For example a company from Saint-Etienne claimed to have been able to reduce mortality from 15% to 5% and morbidity from 50% to 15% by a preconditioning programme

228

Fig. 9. Geographic origins of horses imported into France for slaughtering
(Macchia, 1981).

TABLE 5

ESTIMATION OF THE ANNUAL RABBIT MEAT PRODUCTION IN VARIOUS EUROPEAN COUNTRIES.
NOTE THAT ONLY 55% OF THE FRENCH PRODUCTION IS COMMERCIALISED, WHICH STILL
AMOUNTS TO 93 MILLION RABBITS (SINQUIN, 1970).

Rabbit meat production	
France	220 000 t
Italy	120 000 t
West Germany	25 000 t
UK	15 000 t
Belgium	5 000 t
Netherlands	3 000 t

Transport

- Collection by poulterers
- Transport from maternities
  to fattening units
- Transport to abattoir

realised over a 3-week period and including introduction to the
future feeding rations and sanitary and medical treatments, at
a total cost of 100 F/calf. Judging from these results, it can

only be hoped that other French companies do not hesitate to spend at least the same amount of money in order to get a similar result!

4. Most research has been concerned with only limited aspects of transport stress in farm ânimals. A global consideration of the transport process, including the qualification of animals for transportation, the way they are collected, loaded and unloaded, the design of the vehicles, the number and duration of stops, drinking and feeding facilities during the transit, the qualifications of the staff handling the animals and the relationship of all these factors with the various consequences of transport and not only one aspect such as mortalities or meat quality, has not yet been undertaken. Efforts in this direction are necessary to get a satisfactory evaluation of the transport stress and its impact on animal welfare. Such an objective is certainly not out of reach but would require a close collaboration between the different technical and administrative bodies dealing with transport of farm animals.

ACKNOWLEDGEMENTS

This work has been supported by INRA, Pathologie animale. Appreciation is expressed to Drs. Chauvet, Morisse, Mormède, Mouthon and Soissons and to Mr. Monin for their assistance during the preparation of this review.

REFERENCES

Chauvet, P., 1977. Les veaux: Origines, races, caractéristiques, conséquences. Soc. Française de Buiatrie, pp. 31-36.
Chauvet, P., 1981. Influence du stress avant abattage sur la qualité des carcasses. ULN, rapport n° 163.
Dantzer, R., 1970a. Etude des pertes de poids subies par des porcelets au cours de transports. Ann. Rech. vét., 1, 179-187.
Dantzer, R., 1970b. Etude de l'effet des tranquillisants sur les pertes de poids constatées au cours de transports de porcs en croissance. Colloque sur les transports d'animaux, Institut NOE, pp. 3-5.
Dantzer, R. and Moremède, P., 1979. Le stress en élevage intensif. Masson, Paris.
Dumesnil du Buisson, F. and Signoret, J.P., 1962. Influences de facteurs externes sur le déclenchement de la puberté chez la truie. Ann. Zootechn., 11, 53-59.
Fort, M., 1979. Le ramassage des volailles. ITAVI, Paris.
Gire, P. and Monin, G., 1979. Influences of prednisolone or cortisol injection on muscle glycogenolysis during transport stress in lambs. 25th European Congress of Meat Research Workers, Budapest, pp. 34-37.

Macchia, G., 1981. Rapport sur la mise en vigueur de la Convention europ-
     éenne sur la protection des animaux en transport international: le
     cas des importations françaises de chevaux. Rapport au Conseil de
     l'Europe, Strasbourg.
Micaux, P., 1980. L'Homme et l'Animaux. La Documentation Française, Paris.
Monin, G. and Gire, P., 1979. Influence of injection of alpha- blocking
     and beta- blocking agents on the muscle glycogenolysis during transport
     stress in lambs. 25th European Congress of Meat Research Workers,
     Budapest, pp. 30-33.
Morisse, J.P., Drouin, P., Cotte, J.P. and Huonnic, D., 1981. Influence
     des circuits d'approvisionnement sur les paramètres biochimiques et
     hématologiques du veau. Institut d'Elevage et de Pathologie,
     Ploufragan.
Mormède, P. and Dantzer, R., 1978. Behavioural and pituitary-adrenal
     characteristics of pigs differing by their sensitivity to the malignant
     hyperthermia syndrome induced by halothane anesthesia. II. Pituitary-
     adrenal function. Ann. Rech. vét., 9, 569-576.
Mormède, P., Soissons, J. and Dantzer, R., 1981. Etude des effets biologiques
     du transport et de l'allotement chez le veau. Rapport 8/1.
Mouthon, G., Longin, C. and Magat, A., 1976. Effet du transport sur le pH,
     la pCO$_2$ et la pO$_2$ dans le sang de jeunes bovins de boucherie. Bull.
     Soc. Sci. vét. Méd. comp., 78, 333-336.
Rayssiguier, Y., 1977. Hypomagnesemia resulting from adrenaline infusion
     in ewes: its relation to lipolysis. Horm. met. Res., 9, 309-314.
Sinquin, J.P., 1979. Le lapin: l'economie de la branche, la production,
     le marché, les échanges, in: Les aspects économiques de la production
     du lapin en France, ITAVI, Paris, pp. 1-18.

DISCUSSION

R. Moss *(UK)* I was extremely interested in Dr. Dantzer's remarks about
calves and transport. For the record, all calves that are exported, as far
as we are aware, do not go for direct slaughter, and in any case must spend
a minimum of 10 h in an approved lairage under the supervision of a veterinary officer and be inspected prior to export for their fitness to travel
and their health.

D.B. Stephens *(UK)* We found as Dr. Dantzer has said, and as we have been
discussing, that there are many detectable physiological and psychological
changes which accompany transportation in animals, particularly young ones.
When you have calves produced in the western part of Britain, such as South
Wales, these calves may go to 3, 4 or more markets over a period of 8 - 10
days. What we ought to consider, I think, is how we can determine whether
the animals can adapt to this and secondly how can we diagnose if they have
been subjected to this sort of procedure?

R. Dantzer *(France)* We know what is changing and what is not changing very
well. What we do not know is whether the changes we observe depend on the
extent and duration of the transport conditions. This is what we really need
to know at the moment. This is why the epidemiological approach in France
now is to study two circuits, a long and a short one.

P.V. Tarrant *(Ireland)* I should like to ask you, in relation to your data
on cortisol, whether this might still be a useful indicator of stress, because
the data which you showed us could indicate that the animals were becoming
acclimatised to the simulator and were no longer stressed, in which case the

indicator is a useful one. I think Dr. Kallweit looked at cortisol in cattle
and suggested its usefulness as an indicator of pre-slaughter stress.

R. Dantzer I accept your remark with just one reservation, which is that I
think that cortisol levels are of little value for long term, chronic stress.
There is no problem with short term stress. You can measure the intensity
of the stress by measuring cortisol levels. I am sure that this is so in
pigs, but I cannot say for calves.

P.V. Tarrant When you say chronic long term stress are you talking in hours,
days or longer periods of time?

R. Dantzer Days and weeks.

G. von Mickwitz *(Federal Republic of Germany)* I should like to point out
that it will be necessary to have minimal requirements for the transport of
animals. I have pointed out already that we have made some for the transport
of animals by sea, air, road and rail over the last six to ten years. I think
the best thing would be, if people are interested, to see what other people
have done, compare it and discuss it. If any of the participants are interested I will send them a copy of our minimal requirements.

FINAL DISCUSSION

H.C. Adler *(Denmark)* I am sure you will agree that we have had much information laid before us during these past two days. We have had a great deal of data from scientific investigations as well as results from a wide variety of other forms of observation. We can now discuss welfare aspects of the transport of farm animals on this basis.

Much of the material which has been presented has been provided with other primary aims than the elucidation of welfare. This does not mean, however, that it cannot be used for welfare considerations.

I should now like to provoke the participants by letting them know some of the thoughts which have passed through my mind during these sessions. I have become convinced that with our present knowledge much can be done to increase levels of welfare in transportation. This can be done in a number of ways. Firstly, better design of facilities can be established at the place of origin from which the animals are removed. Secondly, there should be better design of loading equipment, better design of the means of transportation and finally, better design of structures at the point of arrival, particularly for slaughter animals.

I am also convinced that much can be gained by improving the behaviour of those people who work in all stages of transportation. Ways and means to provide improvements on these two points, the better facilities and the better use of those facilities, may vary. I think it could be covered by one phrase: supply of relevant information to the relevant people.

The economic implications of such action could be considered as falling into three categories:

1) Those which can be brought about with no significant cost.

2) Those which may require expenditure.

3) Those which, in connection with the two foregoing categories, may also improve profitability of the animal industry, as well as improving welfare, by reducing losses and improving the quality of animal products.

Having dealt with what I think can be done in the light of present knowledge, I should like to turn to what I see as some of our difficulties. We have seen a lot of figures on physiological, ethological and pathological parameters over these two days. We can draw conclusions from these about their impact on animal welfare. We can use some as merely saying something about animal welfare, because we are not sure, and others cannot be used for animal welfare considerations at the moment.

We have used the word 'stress' a lot, both yesterday and today. We faced difficulties in measuring it, and difficulties in distinguishing between stress and overstress and in evaluating their impact on welfare. I believe that the continuation of relevant research in these fields should be encouraged as it should provide us with more information on these problems.

I also believe that much information can be obtained through further observation of various parts of the transportation procedures as they are carried out in practice.

I am not clear in my mind to what extent this meeting has added to the basis for EC consideration of this problem. It may be that when we see the report of this meeting we may find points which need further elucidation.

In conclusion, I would like to emphasise that EC legislation at present deals only with the international aspect of transport. I should like to finish by putting an impertinent question to my colleagues: is the EC handling of international transport a first step, to be followed by provisions for the transport of animals within member countries, as vaguely indicated in the existing directive? Many more animals are involved in national transport.

J.P. Signoret (France) To my mind it is normal not to have found much material dealing with behaviour in either the papers or the discussion. When we talk about the rearing of animals there is always the possibility of comparing the situation in which we place the animals with the so-called natural situation, whereas transporting animals is, by definition, an abnormal situation. It is absolutely normal, therefore, that in such an abnormal situation we have no reference for what could be termed as normal. We can only rely on physiological indices, and it seems hardly possible to infer from behavioural observations whether or not this is good.

Some animals are transported for slaughter. From the animal welfare point of view, we have to consider this. Up to now, the discussion has dealt mainly with meat quality. This is an important economic factor. Transport under comfortable conditions appears to me to be necessary, but later it is possible that it will be thought that the shorter the interval before slaughter the better, rather than allowing the animal to recover. The animal has been moved into a new social situation and environment, and to have the animal undergoing the process of adaptation is not a behavioural welfare situation, in my opinion.

Finally, I should like to refer to something which strictly speaking is not the subject of this conference but which has to do with welfare. I refer to work by Dr. Guillotin in the eighteenth century, in which he says that his new invention, the guillotine, is designed from a welfare point of view because the condemned man will only feel the breath of fresh air on his neck. This is a good welfare solution.

R. Moss (UK) What I mean by observations of behaviour during transport is more what does an animal prefer to do during transport? Which way does it prefer to orientate itself, how does it relate itself to its companion animals? That is the sort of thing which can influence the design of transport and yet there is a great deal which we do not know and have not observed, e.g. simulators have been discussed but there has been little use of video cameras as yet. With regard to distance is there a limit? We have talked about 14 days being reduced to 7 days by sea. We have talked about the possibility of a maximum distance for animals going to slaughter but are there other factors in relation to distance at which we should be looking? That has not been answered to my satisfaction.

Finally, we have talked about regulations. Do we increase the number of regulations, and find that they are impossible to implement simply because there are so many? As somebody pointed out earlier, are they the right regulations? What was their basis? With regard to immediate slaughter, I think Dr. Lister made the point that lairage was designed as a holding place to allow the line to continue throughout a long period of the day, and that perhaps its use has been extended too far.

P.V. Tarrant (Ireland) I should like to point out that we came to similar conclusions at the end of the previous welfare seminar on dark-cutting in beef, where the participants felt quite strongly that the problem which we were discussing, namely pre-slaughter stress in cattle and the consequent deficiency in meat quality, was largely caused at the pre-slaughter point, and that any regulations or recommendations which existed and which tended to prolong the holding of animals post transport and pre-slaughter were in general detrimental to both animal welfare and meat quality.

It was, of course, very difficult to generalise even within one species because there were different categories of animals and they would

have to be considered separately rather than all together. In that context,
it is even more difficult to generalise across the various species and make
general recommendations of the type made by the European Convention for the
Protection of Animals for Slaughter. These state that animals should be
slaughtered immediately. One of the recommendations of the present meeting
is that in the case of pigs this would not satisfy everybody since for this
species it might be necessary to consider a short holding period.

W. Sybesma *(The Netherlands)* Factors which allow us to assess and record
welfare are important. It is also very important to know which kind of
facilities to use. This has to be based on the indicators we have for
assessing welfare. I think there is an equilibrium between these two factors and between the facilities we use and whether they are sufficiently
adapted to the species we transport. I should like to stress the fact that
we should define certain indicators for the assessment of welfare.

A. Hoogerbrugge *(The Netherlands)* Dr. Signoret made a nice point about
the guillotine, but the condemned person usually was granted some favour
during his last 24 h of life. This contrasts somewhat with farm animal
husbandry!
      We are specialists in this audience, and I agree with Dr. Adler that
we have the opportunity of improving with general regulations and with
specific regulations. I am involved with day-old chicks, but let us say
that if we have a regulation here in Europe relating to the design of chick
boxes, it would be possible to have a very good design and reduce mortality.
It would be good for each country and good for the chicks. There are some
things we can do, and we should do them. We have a general responsibility
to improve the welfare of animals.

J.E. Melville *(Australia)* I should like to make one comment as an outsider
whose principal area of interest has been veterinary public health over the
last eight years. I attend many international conferences of a specialist
nature. We have been talking about transport welfare. I do not think anyone has adequately answered the question, "which is the best time to kill an
animal after arrival at the abattoir?" At a conference on microbiology, the
microbiologists have different ideas from those of you there.
      In food science the food scientists are going to have different ideas.
I think there is a need, in meat science to integrate some of the research
activities which are going on. The industry wants to know, i.e. the producer, the transporter and the slaughterer want to know these answers.
      In our country, as in most countries in the world, the decision is
left to the slaughter plant veterinarian. In many instances he does not
have the information available, he cannot make an objective decision on when
they should be slaughtered or when they should be withheld from slaughter.

H.C. Adler I shall take two conclusions away with me.
      The first is that more welfare in connection with animal transport-
ation can be provided by making better use of present knowledge for such
purposes as improvement of structures and equipment at the place of origin,
improvement of means of transportation, improvement of structures and equipment at the place of arrival and improvement of the function of personnel
involved.
      The second is that research, including all sorts of observations, is
needed which demonstrates the relationship between welfare improvements in
farm animal transportation and the economics of the relevant industries.

LIST OF PARTICIPANTS

BELGIUM

Dr. J.M. Bienfait
Faculté de Médécine Vétérinaire
Rue des Vétérinaires 45
1070 Brussels

Dr. L.O. Fiems
National Institute for Animal
  Nutrition
Scheldeweg 68
9231 Melle - Gontrode

DENMARK

Dr. H.C. Adler
Royal Veterinary and Agricultural
  University
Bülowsvej 13
DK-1870 Copenhagen V

Dr. J.Y. Blom
Institute of Internal Medicine
Royal Veterinary and Agricultural
  University
Bülowsvej 13
DK-1870 Copenhagen V

Dr. Else Hermann
Zoological Laboratory
University of Copenhagen
Universitetsparken 15
DK-2100 Copenhagen

Dr. V. Jensen
National Institute of Animal
  Science
Trollesminde
Roskildevej 48
DK-3500 Hillerod

Dr. Nanny Stenum
Zoological Laboratory
University of Copenhagen
Universitetsparken 15
DK-2100 Copenhagen

FEDERAL REPUBLIC OF GERMANY

Dr. U. Andreae
Institut für Tierzucht und
  Tierverhalten
FAL
Mariensee
D-3057 Neustadt 1

Dr. H. Bogner
Bayer. Landesanstalt f. Tierzucht
D-8011 Grüb 6
Munich

Dr. H.M. Eichinger
Lehrstuhl für Tierzucht
Der Technischen Universität
  München
D-8050 Freising - Weihenstephan

Dr. D.T. Smidt
Institut für Tierzucht und
  Tierverhalten
FAL
Mariensee
D-3057 Neustadt 1

FRANCE

Dr. J.F. Grongnet
ENSAR; Zootechnie
65 rue de Saint-Brieuc
F-35042 Rennes

Dr. P. Le Neindre
INRA
Laboratoire de la Production
  de Viande
CRZV Theix
F-63110 Beaumont

Dr. J.P. Quillet
ITEB
Monvoisin
Cedex 71 bis
35630 Le Rheu

Dr. J.P. Signoret
INRA Comportement Animal
Nouzilly
F-37380 Monnaie

GREECE

Dr. J. Papanicolaou
Ministry of Agriculture
Veterinary Services
Athens

IRELAND

Dr. F.J. Harte
Agricultural Institute
Grange
Dunsany
Co. Meath

Dr. P.J. O'Connor
Deputy Director Veterinary Services
Department of Agriculture
Agriculture House
Kildare Street
Dublin 2

Dr. P.V. Tarrant
Meat Research Department
Agricultural Institute
Dunsinea
Castleknock
Co. Dublin

ITALY

Dr. A. Romita
Istituto Sperimentale
  per la Zootecnia
Via Panvinio 11
Rome

Dr. Marina Verga
Istituto di Zootecnica
Facoltà di Medicina Veterinaria
Università degli Studi di Milano
Via Celoria 10
20133 Milan

LUXEMBOURG

Dr. E. Wagner
Administration des Services
  Techniques de l'Agriculture
16 route d'Esch
Luxembourg

NETHERLANDS

Dr. C. Holzhauer
Gezondheitsdienst Voor Dieren
  in Gelderland
PO Box 10
6880 BD Velt

Dr. L.H. Huisman
Director
Research and Advisory Institute
  for Cattle Husbandry
Runderweg 6
8221 RA Lelystad

Dr. A.A. Jongebreur
Institute of Agricultural Engineering
Postbus 43
6700 AA Wageningen

Dr. G. van Putten
Research Institute for Animal
  Husbandry 'Schoonoord'
PO Box 501
3700 AM Zeist

Dr. N. Steenkamer
Research Institute for Animal
  Husbandry 'Schoonoord'
PO Box 501
3700 AM Zeist

Dr. W. Sybesma
Director Research Institute for
  Animal Husbandry 'Schoonoord'
PO Box 501
3700 AM Zeist

UNITED KINGDOM

Dr. F.R. Bell
Professor of Veterinary Medicine
Royal Veterinary College
University of London
London NW1 OTU

Mr. R. Moss
Animal Health Division
Ministry of Agriculture, Fisheries
  and Food
Hook Rise South
Tolworth
Surrey

Dr. D.B. Stephens
Institute of Animal Physiology
Babraham
Cambridge
CB2 4AT

Dr. A.J. Webster
Professor of Animal Husbandry
University of Bristol School of
  Veterinary Science
Langford
Bristol
BS18 7DY

CEC

  J. Connell
  W. Goldhorn
  J. Wyllie